THE SCIENCE AND CRAFT OF ARTISANAL FOOD

Edited by

MICHAEL H. TUNICK
AND
ANDREW L. WATERHOUSE

THE SCIENCE
AND CRAFT
OF ARTISANAL
FOOD

OXFORD
UNIVERSITY PRESS

OXFORD
UNIVERSITY PRESS

Oxford University Press is a department of the University of Oxford. It furthers
the University's objective of excellence in research, scholarship, and education
by publishing worldwide. Oxford is a registered trade mark of Oxford University
Press in the UK and certain other countries.

Published in the United States of America by Oxford University Press
198 Madison Avenue, New York, NY 10016, United States of America.

© Oxford University Press 2023

CIP data is on file at the Library of Congress

ISBN 978–0–19–093658–7

DOI: 10.1093/oso/9780190936587.001.0001

Printed by Sheridan Books, Inc., United States of America

CONTENTS

List of Contributors   vii

Introduction: Marketing Artisanal Products   1
RICARDO VILLARREAL

1. Beer: Artfully Scientific on Every Level   23
   SCOTT K. UNGERMANN AND CHARLES W. BAMFORTH

2. Artist Winemaker   42
   ANDREW L. WATERHOUSE AND NICK E. GISLASON

3. Artisanal Chocolate   58
   MICHAEL H. TUNICK AND JERRY TOTH

4. Artisanal Coffee   73
   KELLY SANCHEZ, AUSTIN M. MROZ, AND
   CHRISTOPHER H. HENDON

5. Artisanal Cheese   128
   MICHAEL H. TUNICK, SEANA DOUGHTY, AND
   MARÍA PATRICIA CHOMBO MORALES

6. The Chemistry and Flavor of Artisanal Honey   143
   KATIE UHL, ALYSON E. MITCHELL, AND AMINA HARRIS

CONTENTS

7. Industrial and Artisanal Olive Oil   172
   MIKE MADISON

8. Artisanal Fruits and Vegetables   192
   ROSEMARY E. TROUT AND JOSEPH J. TROUT

CONTRIBUTORS

Charles W. Bamforth
Department of Food Science
and Technology, University of
California
Davis, CA

María Patricia Chombo Morales
CIATEJ Unidad Zapopan
Zapopan, Jalisco, México

Seana Doughty
Bleating Heart Cheese
Tomales, CA

Nick E. Gislason
El Retiro Ranch
Napa, CA

Amina Harris
Honey & Pollination Center,
University of California
Davis, CA

Christopher H. Hendon
Department of Chemistry
and Biochemistry, University
of Oregon
Eugene, OR

Mike Madison
Yolo Press Olive Oil
Winters, CA

Alyson E. Mitchell
Department of Food Science
and Technology, University of
California
Davis, CA

Austin M. Mroz
Department of Chemistry,
Imperial College London
London, UK

Kelly Sanchez
Blue Bottle Coffee Company
Oakland, CA

Jerry Toth
To'ak Chocolate
Quito, Ecuador

Joseph J. Trout
Department of Physics,
Stockton University
Galloway, NJ

Rosemary E. Trout
Department of Food &
Hospitality Management,
Drexel University
Philadelphia PA

Michael H. Tunick
Department of Food &
Hospitality Management,
Drexel University
Philadelphia, PA

Katie Uhl
Department of Food Science
and Technology, University of
California
Davis, CA

Scott K. Ungermann
Anchor Brewing Company
San Francisco, CA

Ricardo Villarreal
Marketing Department,
University of San Francisco
San Francisco, CA

Andrew L. Waterhouse
Viticulture and Enology,
University of California
Davis, CA

INTRODUCTION

Marketing Artisanal Products

RICARDO VILLARREAL

To help provide a mental image that describes the environment for artisanal producers and their consumers, consider the following analogy: think of artisanal producers and artisanal consumers as two single people looking for true love based on commitment, understanding, trust, and long-term best interests, *pragmaphiles*.[1] Further, these two pragmaphiles live in a world with many single and married people who are simply looking for passion, lust, and pleasure (*erophiles*[2]). Both our pragmaphiles and erophiles use the word *love*, but pragmaphiles use it to truly express who they are and what they believe, while erophiles use it to for conquests only, with no understanding of what it really

---

[1] *Pragmaphile* is the author's contraction of the ancient Greek word for love based on commitment, understanding, and long-term best interests (*pragma*).

[2] *Erophile* is the author's contraction of the ancient Greek word for love based on passion, lust, and pleasure (*eros*).

Ricardo Villarreal, *Introduction* In: *The Science and Craft of Artisanal Food*. Edited by: Michael H. Tunick and Andrew L. Waterhouse, Oxford University Press. © Oxford University Press 2023.
DOI: 10.1093/oso/9780190936587.003.0001

means to love another. How will two pragmaphiles find each other in such an environment? They will recognize each other by staying true to the definition of love in their hearts. In doing so, they will trust each other through their words, actions, and promises kept. Similarly, artisanal producers can find and keep artisanal consumers through what they say and how they communicate it (words); production processes, ingredients, and distribution methods (actions); and keeping true to the artisanal values (promises kept) that both believe in.

In a marketplace where any brand, artisanal or not, can label and advertise products as artisanal creates a great challenge for truly artisanal producers and consumers looking for such products. In this environment, the meaning of artisanal becomes highly diluted, if not empty. As a result, the word *artisanal* is problematic. It is problematic because there is no formal, legal definition. As a result, the marketplace is replete with products that claim to be artisanal or that have artisanal qualities. For the average consumer, this may cause great confusion. This issue is fundamental to the marketing environment in which artisans operate and consumers shop. Articulating what current artisanal producers do in such a marketplace is an important step toward defending their products and integrity. Unlike cognac and champagne, there is no formal, legal definition to protect artisanal producers and interested consumers. However, a key to success in this environment is to focus on the marketing mix (product, price, place, and promotion—the 4Ps) and other related consumer behavior concepts. Relying on these fundamental, tried-and-true concepts of marketing is analogous to artisans relying on traditional methods of production. That is, there is great value in remembering and following foundational concepts and procedures to achieve the intended outcome.

This chapter will provide a short background on the issue of the term *artisanal* and the problems that have arisen from it. This is followed by a review of the marketing mix and related concepts, with descriptions of how truly artisanal producers can leverage them to identify, attract, and keep consumers who are looking for truly artisanal products. The chapter will close with marketing suggestions for consideration.

Artisanal: What Does It Mean?

The first question is what is an artisan and what is artisanal? According to Merriam-Webster.com, the first recorded use of the noun *artisan* occurred in 1538 (https://www.merriam-webster.com/time-traveler/1538?src=learn-more-timetraveler). The citation, unfortunately, does not provide the context of the recorded use. The provided definitions of artisan are as follows: "1: a worker who practices a trade or handicraft" and "2: a person or company that produces something (such as cheese or wine) in limited quantities often using traditional methods." From artisan comes the adjective *artisanal*. According to the same source, artisanal first appeared in print in 1939, though the context is not given (https://www.merriam-webster.com/time-traveler/1939). The definitions provided are as follows: "1: of, relating to, or characteristic of an artisan" and "2: produced in limited quantities by an artisan through the use of traditional methods; creating a product in limited quantities by traditional methods."

There are also definitions created by practitioners and organizations, with many definitions having elements of those found in a dictionary. A few examples are provided that identify important aspects of the term *artisanal*. In 2005, James Mellgren, in *Gourmet Retailer*, provided the following definition of artisanal: "products that are made by hand, usually in small batches, usually adhering to age-old traditions, even though at times the products themselves can be innovative. Typically, these products reflect one producer's vision." In a 2013 dissertation, Dr. Jenifer Buckley provided a definition of artisanship as being of "small or medium scale, [emphasizing] manual techniques and close producer involvement, and [accommodating] variability in products and processes" (p. 58). The Center for Urban Education about Sustainable Agriculture (2006) defines artisanal as "a food product . . . made by a skilled craftsman, using high-quality ingredients and a mastered, often traditional technique." Suzanne Cope (2013) adds that an artisanal product is "made in small batches and with ingredients that are sustainably sourced" (p. 4). A final definition comes from the *Final Report of the International Symposium on Crafts and the International Market: Trade and Customs Codification* (UNESCO and International Trade Centre, 1997):

> Products that are produced by artisans, either completely by hand or with the help of hand-tools or even mechanical means, as long as the direct manual contribution of the artisan remains the most substantial component of the finished product. . . . The special nature of artisanal products derives from their distinctive features, which can be utilitarian, aesthetic, artistic, creative, culturally attached, decorative, functional, traditional, religiously and socially symbolic and significant. (p. 6)

The variation in the definitions suggests a lack of consensus and adoption of a single, formal, universally accepted definition. Without a formal definition, brands are free to create their own or to use existing ones as they see fit. Indeed, many national and global brands use some form of *artisanal* indiscriminately.

Would the creation of a universal, legal definition of artisanal solve the problem? Although the process of creating such a definition would undoubtedly be arduous, doing so could create a double-edged blade, providing benefits and restrictions for both producers and consumers. A benefit for producers, for example, is that it would identify, and separate, truly artisanal producers from those who currently use the term solely as a marketing ploy. This could be important as global brands, including McDonald's (artisan grilled chicken), Domino's Pizza (four types of artisan pizzas), and Dunkin' Donuts (artisan bagels), have positioned some of their mass-produced products as artisanal. For consumers, a legal definition could help in their search for artisans and artisanal products. For example, consumers would be able to reduce the time they would spend distinguishing truly artisanal producers from those who are not. A negative outcome of a legal definition may be the implementation of restrictions that could be detrimental to producers. As Buckley (2013) noted, a standardized definition many require regulations, such as for food safety, that may well make survival difficult for artisanal producers. For example, Buckley (2013) conducted a study on cheese and jam artisans in Michigan and found that regulating food safety negatively influenced the artisanal production processes and added other burdensome restrictions. Such regulations increase the cost of doing business, which leads to price increases for consumers and a likely drop in demand. These findings suggest that protecting the term *artisanal* may help in one regard while creating negative outcomes in another.

Undoubtedly, the misuse of the term *artisanal* is a major problem for producers and consumers. How can small artisanal producers compete against national and global brands? A first step is for artisanal producers to identify the variables they can't control from those they can control. Clearly, artisanal producers can't control the use of the term in the marketplace. Although this is a problem, artisanal producers have control over a very powerful set of marketing tools that can be used to position themselves in the minds of consumers. Presented below, these tools are used to execute marketing strategies relative to complexities in the artisanal goods marketplace. With the ability to control them, producers can differentiate their goods from those that are artisanal in name only. These fundamental tools give artisanal producers the power of David in the face of Goliath.

The Marketing Mix

The *marketing mix* refers to the tools that a company has the power to adjust to either take advantage of an opportunity or mitigate threats in the marketplace. The 4Ps stands for product, price, place (distribution), and promotion (marketing communications). *Product* refers to all the goods or services offered by a business. Price is the amount of money one must pay in exchange for a business entity's product or service. *Place* refers to distribution. It includes all the methods by which and the locations where a consumer can purchase a company's product or service. *Promotion* refers to a company's use of communications to make consumers aware of their product or service and benefits. An important aspect of the 4Ps is to understand that collectively they help define, articulate, and reflect a brand's positioning in the marketplace. As simple as these may be, they are the most important tools that brands use to execute marketing strategies.

Brand Positioning and the 4Ps

Brand positioning is an important concept closely related to the 4Ps. It is defined as "the act of designing the company offer and image so that it occupies a distinct and valued place in the minds of target customers" (Kotler and Armstrong, 2013). Brand positioning is implicitly relative to competing brands. That is, a brand can position itself as being part of a particular product category by identifying points of parity (e.g., Audi, BMW, and Mercedes all share German engineering and racing history). Additionally, brands may distance their products from a product category by identifying points of differentiation (e.g., Audi in terms of design as compared to BMW and Mercedes).

Kotler and Armstrong's definition of positioning suggests two primary components, one product-related and the other consumer-related. The product component is related to a unique attribute or benefit that the product possesses. The second component refers to the brand image that resides in the minds of consumers (i.e., brand positioning). A product's unique attribute or benefit is a brand's way of telling consumers why their product is different (i.e., better) and explaining the benefit(s) of this difference. The second component relates to the positive thoughts and images about a brand that reside in the minds of the consumer.

How does a consumer develop thoughts or images about a brand? It's no accident. Positioning is the product of everything a brand does and communicates to consumers. Positioning begins to take hold when there is overlap between what a brand says a consumer will experience with their product and a consumer's actual experience with the product. For example, let us say an artisanal brand positions itself as "an authentic, traditional maker of product X." Further, assume the benefit of this brand is the use of traditional methods of production and the quality of ingredients, resulting in a delicious product that, upon consuming, has a great emotional impact on the consumer. This fictitious positioning has implications for the 4Ps. For example, the product must be made of quality ingredients and have an amazing taste. The price, too, must reinforce this positioning. What would one expect to pay for a product made as described in the positioning? It would certainly cost more than a similar industrially created product. In terms of place (distribution), where would one expect to purchase such a quality product? Uniqueness of product implies a selective distribution approach, only sold in a limited number of exclusive locations as opposed to an intensive distribution (being sold in all mass market locations). If consumers can purchase "product X" at the production facility, what visual, auditory, and sensory experiences does the positioning suggest a consumer would have? Finally, in terms of promotion, all communications, regardless of medium and goal, must reflect the brand's actual and image-based aspects alluded to in the positioning. It is crucial to note, however, that what is important or unique from the brand's perspective must also be important or unique from the consumer's perspective. In other words, if consumers do not believe that a brand's benefit or uniqueness is important, then the brand has no real chance of convincing consumers to try the product.

Consumer Behavior

The consumer behavior process (CBP) is conceptualized as a series of steps that consumers go through on their purchase journey.

The first step of the CBP is *need recognition*. In this step a consumer first determines there is a difference between their desired state and their actual state. In other words, a consumer believes they are experiencing a consumption-based need that a brand might be able to satisfy. The second step is *information search*. In this step, the consumer must determine which brand or brands might have the best product or service to satisfy their consumption need. To do this, consumers may use one or both search options available to them: internal search or external search. *Internal search* is where a consumer scans their mental memory banks for prior experience with the current need or a situation similar to the current need. This search provides relevant information on how to solve their current consumption problem. *External search* includes, but is not limited to, personal sources (family, friends, neighbors, acquaintances), commercial sources (advertising, salespeople, brand websites, packaging, displays), and experiential sources (handling, examining, using the product) (Kotler and Armstrong, 2014).

Regardless of whether internal, external, or both searches are used, consumers now have a list of brands they believe may address their consumption need. The third step is the *evaluation of alternatives*. After a consumer determines a list of alternative offerings, they must then evaluate the brands in their list of viable alternatives. The penultimate step is the *purchase decision*. After evaluating their brand list, the consumer then makes a purchase from the brand they feel will best address their issue. Finally, the CBP ends with the *post-purchase behavior/evaluation*, where the consumer determines if the selected brand met their expectations or not.

Brand awareness is an important component of the CBP's internal search step. Brand awareness is simply what the term implies: Are consumers aware of a given brand? If consumers are unaware of a brand, there is no chance the brand will come up in a consumer's internal search. In other words, a consumer cannot consider an unknown brand for purchase. In such a case, a brand with low awareness can only hope that an external search brings their name up for a consumer's consideration.

The CBP outlines the major steps that consumers go through when making a consumption decision. It is important to note that the process can be circumvented. Branding allows consumers the ability to circumvent the CBP as brand image and product experience can act as a shortcut through these steps. That is, a positive experience with a product creates a positive memory that can take a consumer from need recognition directly to product purchase.

The CBP is useful in explaining the steps a consumer may follow in finding a product that meets their consumption need. However, it does not provide any information as to why consumers buy what they do. Understanding the reason(s) consumers make the purchases they do is just as important for brands.

Motivation

Motivation is the concept that helps marketers understand why consumers buy what they do. This understanding is integral to determining the consumer needs a product can meet or if a product can be repositioned to meet other needs. Although there are many needs that impact consumer behavior in general and motivation in particular, the following may be more pertinent for artisanal consumers and those consumers who may be persuaded to consider artisanal goods. The two components are *relationship with a product* and *consumer needs*.

RELATIONSHIP WITH A PRODUCT

For artisanal producers, the most relevant concepts related to a consumer's relationship with a product are self-concept, love, and interdependence. Self-concept suggests that consumers use products that help them define, reinforce, and express to others their sense of identity. For example, artisanal products allow consumers to reinforce their self-image, convey meaning to others about the way the consumer sees themselves, and tell others how they want to be seen. A brand that allows consumers to meet these motivational needs helps create an emotional bond between the consumer, the product, and the brand. The consumer's affective response is the heart of the "love" component of a consumer's relationship with a product. When consumers find a product and brand that allows them to meet their consumption needs and their self-concept, the product becomes integral to their daily routine. This relationship becomes integral to the concept of brand loyalty.

CONSUMER NEEDS

The need for affiliation and that for power are components of the "consumer needs"–based reasons that consumers buy what they do. Arguably, these needs have some conceptual connection to those of the "relationship with a product" aspect of consumer behavior. The difference, however, may be more psychologically grounded. For example, the need for affiliation suggests that consumers purchase products primarily to identify themselves as members of a particular group regardless of other product-related variables such as price or travel time to purchase. This suggests that practical product benefits may be less important than the identity aspect of using the product. For example, a consumer with a great affinity for local artisans and their products is willing to pay more for products and travel farther rather than purchasing an internationally sourced, fair trade, artisan product. The second consumer need, the need for power, has to do with one's desire to control one's environment. The key here is the "desire" to control, suggesting that the purchase of artisanal products may make consumers feel they are doing their best to manage their local environment. This could include purchasing from local artisans, who are part of the local environment and who use locally sourced ingredients.

Cleary, consumers' purchase behavior suggests that consumption is more than meeting a need; it is making a statement. And the extent to which a brand helps consumers define and express themselves suggests how a brand might fit into a consumer's life beyond simply meeting a need.

Segmentation, Targeting, and Positioning

Segmentation, targeting, and positioning (STP) are other fundamental marketing tools. These three tools are interrelated and suggest an ordered process. Segmentation starts with taking an entire market and segmenting it. The next step is to select one or more viable segments to target. The final step is to position the product within the viable target segment(s). As Kotler and Armstrong (2014) state, "[t]here is no single way to segment a market. A marketer has to try different segmentation variables, alone and in combination, to find the best way to view market structure" (p. 193).

As noted above, there are many variables that can be used to segment a market. The major variables, however, include geographic, demographic, psychographic, and behavioral segmentation. Of these variables, psychographic may be considered the most relevant for artisanal products. Psychographics is a concept that is comprised of multiple, higher-level concepts, each of which is defined by multiple variables. The higher-level concepts include social class, lifestyle, and personality characteristics. Social class is usually a concept defined by education level and income level; lifestyle is typically defined as the activities, interests, and opinions that define how one lives. The usefulness of psychographics is that it suggests that consumers with similarities on one or more of these variables may have similar responses to a brand's use of the 4Ps.

Once segments are determined, the next step is to assess their relative viability. That is, is a segment realistically viable? To help answer this important question, marketers collect the data needed to answer the five requirements for effective segmentation: measurable, accessible, substantial, differentiable, and actionable. *Measurable* means that a marketer must be able to measure the size of the segment and the segment's purchasing power. *Accessible* means that a segment can be reached in terms of distribution and communications. So, one must know not only how many are in a segment and what their purchasing power is but also where they live geographically and which media they use for communications. *Substantial* suggests that the segment is large enough or profitable enough to warrant targeting. In other words, are there enough consumers in the segment, and do they have the purchasing power to warrant targeting? *Differentiable* suggests that when multiple segments are identified, each segment must behave differently from every other with changes in the marketing mix. In other words, multiple segments should not equally respond to a single marketing mix because if they do, then they are not actually different segments. *Actionable* suggests that a brand has the ability and resources to effectively attract and serve a segment or multiple segments.

Targeting

Once a segment(s) passes the viability criteria, brands can focus on targeting. There are three major targeting strategies: undifferentiated, differentiated, and concentrated. *Undifferentiated targeting* is essentially mass marketing. This strategy implies that no real segmentation of a market has been done or that the identified segments did not meet the criteria for effective segmentation. This type of targeting suggests that a product can meet all needs for all consumers using a single marketing mix. *Differentiated targeting* suggests that a brand has identified multiple segments that meet the criteria for effective segmentation and decided to create unique marketing mixes for each segment. In this case, at least one of the 4Ps is different between each segment, and each segment has a unique reaction to the marketing mix. *Concentrated targeting* is the selection of only one of multiple, viable segments. For example, of three viable segments, a brand will target the one that it feels can most successfully meet the needs of consumers at a profit.

Positioning

Positioning in this context relates to a product and not the umbrella brand that makes the product. For example, an Audi Q5 has its own product-specific positioning of sport, luxury utility, while at the same time reflecting and reinforcing Audi's umbrella brand positioning of luxury, style, and technology. In the context of segmentation and targeting, positioning is related to the specific and unique product attributes that consumers hold in their minds in relation to competing products. An obvious question is, how do consumers create an image of a brand? Brand image is built and reinforced through the 4Ps and positioning. It is no accident what a consumer thinks about a brand as the brand image and positioning are reflected and reinforced in the product, price, place, and promotion of a brand's product.

Integrated Marketing Communications

Marketing communications is an important aspect of STP. Integrated marketing communications (IMC) is defined as the "coordination of the promotional mix elements . . . with each other and with the other elements of a brands' marketing mixes . . . such that all elements speak with one voice" (Shimp & Andrews, 2013, p. 4). IMC should happen at both the brand level and the product level. In addition, all product-level communications should reinforce the overall brand positioning. One of the fundamental reasons that a brand should speak with a single voice across all media is to reinforce the brand messaging and image. Any inconsistencies in messaging may lead to consumer confusion.

As noted earlier, positioning resides in the minds of consumers. Consumers develop their ideas about a brand through word of mouth, which a brand cannot directly manage; marketing communications; and experience with the brand. Although a brand cannot directly manage word of mouth, managing marketing communications and brand experience provides an indirect way to impact word of mouth.

It is also important to note that consistency in messaging does not only relate to explicit words use in media communications. That is, all formal communication messages about an artisanal product may lead to expectations regarding the actual product. This is important for artisanal producers to note as many artisanal products are highly sensory in nature (touch, sight, sound, smell, and taste). In terms of positioning and messaging, these senses may be just as important as the actual message in terms of brand image and positioning. It's important that a brand ensures that it can meet any expectations consumers may create because of exposure to its messaging.

Possible Use of Marketing Tools for Artisanal Producers

This chapter has so far laid out some of the fundamental marketing and branding options that artisanal producers can use to compete in a product category that is defined by a term that is used indiscriminately. Although this is a problem, it also provides an opportunity. The opportunity is that artisanal brands can self-define what *artisanal* is, why it is important, and why customers will like it. Whatever the specific definition of artisanal a producer develops, they must ensure that it is important to consumers, evident in all aspects of the 4Ps, and consistently messaged in all communications.

4Ps, STP, and Branding

As described at the beginning of this chapter, a main problem for truly artisanal producers is that they compete with truly non-artisanal producers, including national and global brands. This is largely due to the use and misuse of the term *artisanal*. Despite this, truly artisanal producers do have an opportunity. The opportunity comes from their power to define what artisanal is, what it means to them, and why it is important and to consistently convey this message to consumers through the 4Ps and branding.

In terms of the 4Ps and STP, one may ask, what comes first? This is a good question. The answer is to start with what you create, how you create it, why it is unique, and why consumers will like it. With these basic questions answered, one can almost simultaneously work on branding and positioning. The attributes of a brand and its positioning can then be integrated into the 4Ps. Another important thing to remember is that branding and positioning are not static. Product categories and consumers are always changing, and primary marketing research is fundamental to identifying changes. Primary marketing research is needed to track explicit customer wants and needs and to identify latent customer wants and needs. Any changes in consumer behavior may have either a positive or a negative impact on the artisan, their products, and current positioning.

Possible Marketing Responses for Artisanal Producers

In the current environment, each artisanal producer is doing what they can to survive in their own unique context. Right now, there are untold numbers of artisanal producers facing the same basic problems but each fending for themselves. This fragmentation indicates that producers would benefit by combining their efforts to manage the negative issues they share. In short, this current environment suggests a need for basic unification through an organized body to help create awareness and to protect individual artisanal producers. For example, for areas with a good number of artisanal producers, it might be worth considering the creation of something like a medieval craft guild. The most relevant function of a medieval guild was that it was responsible for creating and monitoring the standards related to the crafts it represented. All local-level guilds could then collaborate to create a regional guild. Logically, all regional guilds could then collaborate to create a national guild. This type of hierarchy could increase awareness of the issues noted earlier by creating a definition(s) for what an artisan is and what defines their products. The national guild could also warn consumers of the misuse of the relevant terms, building trust in the guild and its members. The work of a national guild and its logo could help build awareness on a national level, which would then benefit guild members at the regional and local levels.

Artisanal Taxonomy

In the current marketing context for artisanal producers, one thing that can be done by artisans is to create a taxonomy of the components that define artisanal products. This taxonomy can then be used to provide evidence of the "artisanality" of a producer and their product(s). The components presented here come from the varied definitions of artisanal presented earlier. It is important to note, however, that each definition treats the measurement of artisanal as a discrete variable. That is, one either is or is not an artisanal producer. It might be more useful to think of "artisanality" on a continuum that ranges from artisanal to not artisanal. The following components are integral to defining artisanal products: production process (traditional–modern), method of production (hand–mechanized), output (small–large), distribution (exclusive–intensive), ingredients (minimally processed–highly processed), and artisan involvement (100%–0%). Based on these components, I present the following *Artisanality Measure*. Each component is individually evaluated with a semantic differential scale using appropriate anchoring categories. For example, *production type* refers to the process by which goods are made. Based on definitions, traditional production methods are considered artisanal, so the scale has *traditional* on one end and *modern* on the other. For production method, the anchors are *hand* and *machine*, and so on. A full articulation of the Artisanality Measure is presented in Figure I.1.

The Artisanality Measure can have important, multiple benefits. The goal of the measure is to help define an artisanal product for consumers. As simple as this may sound, defining artisanality can impact consumer attitudes. Attitudes are comprised of the

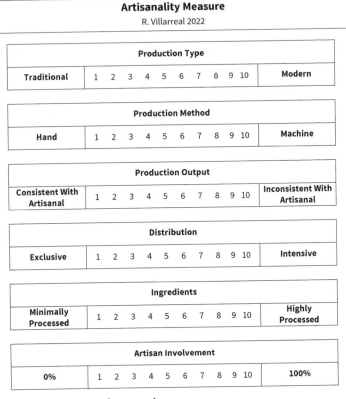

Figure I.1 Measures of artisanality

following three components: beliefs, feelings, and behaviors. *Beliefs* are what consumers currently have in mind for a given product or brand. Beliefs can come from many sources, including information from brands. *Feelings*, as the term suggests, refers to how a consumer feels about a given product or brand, while *behaviors* refers to how a consumer will act toward a product or brand. Importantly, these components are ordered, such that beliefs about a product or brand can impact how a consumer feels about a product or brand, and these feelings can then impact a consumer behavior toward a product or brand. The artisanality measure can help artisans impact consumer behavior by providing objective information that can ultimately have a positive impact on consumer behavior.

Conclusion

In marketing, words have meaning. *Artisanal* is one such word as it sends a positive signal to consumers about what they can expect in terms of the product and the artisan who makes it. Unfortunately, the word has been diluted because national and global brands use it to label their large-scale, highly processed, and mechanized products. As problematic as this is, it is important to note that the battle against the misuse of the term does not start on store shelves. Rather, it starts in the minds of consumers. Arming consumers with truthful information about what is and is not an artisanal product is an important first step in protecting the word. Using an optimal balance between the tools and concepts presented in this chapter, artisanal producers can define and position their products to protect themselves while attracting consumers. The right balance between the tools is critical to creating and maintaining a successful marketing strategy that can create customer value and meaningful, long-term connections between consumers and artisanal producers.

References

Buckley, J. A. "*Artisan* Food Processing *and* Food Safety Regulation *in Michigan: An* Actor-Network Study *of* Interactions, Interests, *and* Fluid Boundaries." PhD diss., Michigan State University, 2013. https://d.lib.msu.edu/etd/2204

Cope, S. "Artisanal Food Production and Craft." In *Encyclopedia of Food and Agricultural Ethics*, edited by P. B. Thompson and D. M. Kaplan, 1–7. Dordrecht, the Netherlands: Springer Science+Business Media, 2013.

Foodwise. "Artisanal defined." October 6, 2006. https://foodwise.org/artic les/artisanal-defined/.

Kotler, P., and G. Armstrong. *Principles of Marketing*. Upper Saddle River, NJ: Pearson Prentice Hall, 2013.

Mellgren, James. "Down to Earth: Merchandising Artisanal Foods." *Gourmet Retailer* 26, no. 12 (2005): 38–40.

Shimp, T. A., and J. C. Andrews. *Advertising, Promotion, and Other Aspects of Integrated Marketing Communications*. 9th edition, Cengage Learning, Independence, KY, 2013.

UNESCO and International Trade Centre. *Final report of the International Symposium on Crafts and the International Market: Trade and Customs Codification*. Paris: UNESCO; Geneva: International Trade Centre, 1997.

1

BEER

Artfully Scientific on Every Level

SCOTT K. UNGERMANN AND CHARLES W. BAMFORTH

Does Size Matter?

The word *craft* has come into popular parlance in the world of brewing, just as it has in other processes and industries. The challenge comes in trying to define it. Where does craft end and *industrial* begin (Bamforth, 2020)?

The Brewers Association (BA; www.brewersassociation.org), the body that was established to promote and protect the interests of "smaller" brewing companies, defines a craft brewing company as being "small and independent." It is at its website that you can unearth the magnitude of the beer sector in the United States, exemplified by the data in Table 1.1. Craft had 13.6% of the market in 2019, and imports had 19.0%, with the remainder of the sales being from the larger brewing entities.

What exactly is "small" in the eyes of the BA? This is identified as an annual production of 6 million barrels of beer or less, which

Scott K. Ungermann and Charles W. Bamforth, *Beer* In: *The Science and Craft of Artisanal Food*. Edited by: Michael H. Tunick and Andrew L. Waterhouse, Oxford University Press. © Oxford University Press 2023. DOI: 10.1093/oso/9780190936587.003.0002

Table 1.1 Recent US Brewery Count

	2015	2016	2017	2018	2019
Craft	4,803	5,713	6,661	7,594	8,275
Large (non-craft)	44	67	106	104	111

Note: There were some 1,800 breweries in the United States in 2012.
Source: www.brewersassociation.org/statistics-and-data/national-beer-stats/.

approximates to the annual output of Denmark. It can hardly be said to be a drop in the ocean.

By "independent," the BA determines that less than 25% of the company can be owned or controlled by an entity within the alcoholic beverage industry that is not a craft brewer. This means a company such as Anheuser-Busch InBev, Asahi, or Heineken. Thus, when the last of these purchased Lagunitas from Petaluma, California, the latter overnight was forced out of the "club." Irrespective of the extent to which a larger company leaves an acquisition to its own devices and approaches, the bought-out concern is no longer "craft."

The BA says that "craft beer is generally made with traditional ingredients like malted barley; interesting and sometimes non-traditional ingredients are often added for distinctiveness." (Brewers Association, 2023). There is a substantial element of subjectivity in this statement. To illustrate, a grist material such as rice, used for more than a century alongside malted barley in the production of Budweiser, is generally perceived as not being "craft"—so, too, corn and corn syrup. However, the candi sugar (derived from beet) used by Trappist brewers is applauded. "Distinctiveness"? Presumably that refers to components such as bull testicles, chili peppers, and pizza crust in "craft" brews? And many a "craft" brewer (like many of the bigger companies) is busily producing drinks such as alcoholic seltzers, which can hardly be considered commensurate with the values espoused by the BA.

The BA does a great job. So, too, do the Institute of Brewing & Distilling (www.ibd.org.uk), the Master Brewers Association of the Americas (www.mbaa.com), and the American Society of Brewing Chemists (www.asbcnet.org), organizations within which the technical people from breweries and related companies come together. There is no size discrimination in these societies. Membership is primarily at the level of the individual, and anyone who belongs to a brewing company (or related entity) may join, irrespective of the volumes of beer produced by their employer. Thus, the brewmaster of a multimillion-dollar brewery within a global company will be rubbing shoulders with a hopeful newcomer who may be brewing a few gallons a month. Yes, the former is managing an operation that is controlled by intricate computerized systems, employing sensors and automation, and taking advantage of detailed market information to dictate production planning. The latter, meanwhile, has large boots and a paddle and is dependent on primitive analytical tools and, as often as not, on their senses and intuition. However, they are both brewing beer using the same fundamental processes. And they share experiences.

A craft brewer is surely someone who is knowledgeable and talented, thereby able to brew beer with skill to delight the customer on whatever scale. When one hires craftspeople to do specialized jobs around the home (electricity, plumbing, and so on), one is

SCOTT K. UNGERMANN AND CHARLES W. BAMFORTH

looking for their abilities to deliver and not considering the size or nature of the company that employs them, although it is certainly true that many of us prefer to support the smaller businessperson from the neighborhood.

Is the local weekly newspaper a better or worse publication than the *New York Times* or the *Wall Street Journal*? Yes, they may have different functions—local information versus global perspective—but it is surely not credible to simply conclude that size is somehow to the detriment of quality.

Perhaps any differentiation between the larger and smaller brewing companies arises when one considers business practices, including sales and marketing. The larger concerns have undisguised "clout." They are in a far better place to "call the shots" with the distribution companies in the three-tier system that comprises a chain of producer, distributor, and retailer. The bigger brewing companies can negotiate advantageous purchasing contracts and put their thumb on the scales of availability. And they can afford Super Bowl advertisements.

Brewing Technology

To reiterate the point, however, the shape of the brewing process is the same at all levels. The same unit operations, which are captured in Table 1.2, are in place. Table 1.2 highlights what differences may exist in the individual stages related to size, and it is certainly the case that there will be a far greater investment on a larger scale.

26

Table 1.2 The Unit Operations of Brewing

Operation	Purpose	Likely difference between a brewery producing 1 million barrels of beer per annum and one producing 1 thousand barrels of beer per annum
Raw material storage	Storage of inventory of malts, hops, etc.	Much greater capacity for the larger companies. Maybe a larger stockholding, purchased at more competitive rates. Silos versus sacks.
Milling	Grinding of malt to facilitate efficient extraction	Simple two-roll mill on the small scale or even purchase of pre-milled malt. Six-roll mills or hammer mills on larger scale, together with magnets, destoners, pneumatic handling, weighers.
Mashing	Mixing of milled grist with hot water to allow enzymic breakdown of starch to fermentable sugars	Likelier use of adjuncts on a very large scale, demanding cereal cookers. Increased likelihood of manual mixing on smallest of scales.
Wort separation	Separation of liquid extract (sweet wort) from spent grains	Increased likelihood of a single vessel for mashing and wort collection on smaller scale, followed by manual removal of spent grains from vessel. More automation on larger scale, together with possible online monitoring of differential pressures, specific gravity, clarity. Greater likelihood of use of mash filters on larger scale.
Boiling	Boiling of wort with hops to extract bitterness, sterilize wort, precipitate protein, remove unwanted flavors, etc.	Increased use of modern heating strategies (e.g., calandria) or latest boiling approaches (e.g., incorporating volatile stripping technology) on larger scale.
Wort clarification	Removal of insoluble materials from the bittered wort	Usually a whirlpool on all scales, though more likely to be a combined kettle and whirlpool on smaller scale.

(continued)

Table 1.2 Continued

Operation	Purpose	Likely difference between a brewery producing 1 million barrels of beer per annum and one producing 1 thousand barrels of beer per annum
Wort cooling	Chilling of boiled wort to fermentation temperature	Usually involves a heat exchanger system on anything other than smallest scale, where cooling may be simply allowing the wort to slowly cool in an open tank, perhaps a coolship.
Fermentation	Use of *Saccharomyces* to convert pitching wort into beer	Likelier use of open fermenters as scale decreases.
Yeast handling	Management of yeast cultures and stocks to ensure hygienic control of yeast used for fermentation	Increased likelihood of employment of dried yeast on smaller scale. Larger breweries will have more sophisticated yeast handling strategies, including hygienic storage facilities, possible use of acid washing, maybe in-line yeast monitoring devices. Numbers of separate strains generally in proportion to size of brewery. Larger companies will have yeast propagation systems.
Maturation	Refinement of beer flavor following fermentation, especially removal of unwanted material such as vicinal diketones and acetaldehyde	Usually more sophisticated approaches on a larger scale, although the principles remain the same on all levels.
Downstream operations (clarification, stabilization)	Removal of unwanted materials, notably those leading to turbidity in those beers that are intended to remain "bright"	If there is any clarification on a small scale, it may be little more than natural sedimentation and possibly simple sheet filtration. On a larger scale usually powder-based or crossflow filtration, use of adsorbents such as silica or polyvinylpolypyrrolidone. Centrifuges.

Term	Description	Details
Packaging	Filling of kegs, bottles, cans	On a small scale maybe no more than a pipe leading from a serving tank in the brewery into the bar, front of house. On the larger scale, tremendous automation and filling strategies capable of highly consistent fills and minimum oxygen levels. Small-scale filling of bottles or cans very manual and labor-intensive.
Warehousing	Holding of beer prior to distribution	Large stockholdings in ideal dedicated warehouses with optimum conditions on a large scale. Minimum stock holding on the smaller scale in whatever space is available.
Distribution and retail	Shipping of beer to holding warehouses and to trade	Fleets of trucks and railcars on a large scale, possibly refrigerated to support the shelf life of the beer. At the other extreme there may be little, if any, movement of beer from the brewery, and then it will be relatively local.
Co-products	Notably spent grains, surplus yeast, carbon dioxide	Pneumatic handling of grain on larger scale, together with fleets of trucks taking the grains away to handlers. On the smallest scales more likely to be a deal with a local farmer coming to collect grain. Yeast also shipped out to processors or farmers on large scale but more likely to go to drain on smaller scale.
Cleaning	Efficient cleaning of all tanks, piping, etc.	Sophisticated cleaning-in-place systems on larger scale. Manual on smallest scale.
Quality assurance and control	Systems for monitoring and ensuring product excellence	On the larger scale a dedicated department that implements a quality program with documentation, standard operating procedures, analyses at all stages, audits, detailed microbiological strategies, etc. On a smaller scale a far simpler approach with minimal instrumentation and measurements.

The Brewer's Goals

When considering the differences between artisanal or "craft" beer and larger-scale operations (sometimes unreasonably called "industrial" beer), it is important to consider the goals of the brewer. In most cases, whether the beer was originally crafted by a 19th-century industrialist brewer or a modern-day craft brewer, the endeavor and the end goal were likely much the same. The brewer has always desired to produce a beer that is delicious and enjoyable for personal drinking and sharing with friends and family as well as the general public. The crafting of a beer recipe depends upon many factors that include, but are not limited to, availability of raw materials (hops, malt, etc.), chosen yeast strains, available brewing equipment, processing ability, and availability of cold storage. Many of these are outlined in Table 1.2. One cannot craft a complete recipe without knowing all the processing constraints of the brewery in which that beer is to be produced. Brewing recipes do not just include a list of materials and a set of times and temperatures. There is much more to consider if we are to look at the complete picture.

Scaling Things Up

To illustrate the levels of complexity and craftsmanship in developing a beer, it is a useful exercise to consider the actions and plans that need to be undertaken in scaling up a beer recipe. At first this will be from a homebrew to a brewpub and then, further yet, from a brewpub to a large production operation. As we look at these points of differentiation and the decisions that need to be made along the way, it will lead to a comparison of scale and what that means to the ultimate outcome—which is (hopefully) a delicious glass of beer, properly poured and served in a clean glass with a nice head of foam.

Starting with a basic homebrew recipe, let us assume that our homebrewer has an excellent recipe for a pale ale and that our brewer is ready to take that formula to the next level and scale it up to a brewpub. For the sake of discussion, our brewer has a simple recipe that delivers a nice refreshing pale at 5.5% alcohol by volume, 35 international bitterness units, and a color of 5.5 Lovibond. (See Bamforth [2006] for an explanation of these terms.) The flavor of the beer is described as slightly estery, pleasantly hoppy, balanced, and crisp. To achieve final gravity in this beer, the homebrewer conducts fermentation in a glass carboy at approximately 68°F (20°C) for about a week. The beer is then racked into a keg, where it is cooled and carbonated. This is a reliable recipe for our homebrewer. The yeast selected performs repeatedly well, and the beer is reliably delicious. Now let us assume that this homebrewer somehow has secured the funding and the audacity to open a brewpub (yet another!) to serve this pale ale to the thirsty public. Our brewer clearly will have to have at least seven other recipes to start said brewpub, but for now let's focus on just this one.

With funding in hand, our brewer now must design, purchase, or build a brewery and a process to go along with all these recipes—but mainly to enable the production of this pale ale because it will be the flagship. First, we must consider the processing equipment upon which this recipe was born. Whether it was a stovetop mash followed by a home-constructed lauter bucket into a gas-fired kettle or some more extravagant homebrew kit such as the

SCOTT K. UNGERMANN AND CHARLES W. BAMFORTH

SABCO BrewMagic (www.brewmagic.com) in that homebrewer's garage, that equipment had something to say about how the beer was made and how the final beer looked and tasted. The properties of mashing and the outcome depend not just upon materials, time, and temperature but also upon degree of milling, the water-to-grain ratio, heat transfer coefficients, and many other factors. The temperatures achieved and the time taken on each process stage have obvious impacts on the outcome, as does the evaporation achieved during the boil—indeed, everything else from start to finish in the operation. All these factors are dependent on the nature and configuration of the brewing equipment—the materials of construction, the heating elements, and the level of automation and control are all intertwined into defining the system on which the brewer works.

When we look at fermentation, the decisions made become even more complicated. The yeast selection process is as difficult and unpredictable as any other but one that must be considered before looking at fermenter shape and size. Yeast handling equipment and techniques also play a very important role in the ability of the brewer to make repeatedly good beers.

Since our homebrew produced a nice estery beer with no notable defects, we are looking for the same outcome once we scale up to a larger operation. The design of the fermentation process is crucial to producing the same beer flavor that the brewer intended. Height-to-diameter ratios of the fermenters matter in this instance; the amount of back pressure on the vessel will impact yeast performance, and the amount of liquid column pressure on the yeast will impact the flavors and aromatics produced during fermentation. Temperature control, or the lack thereof, also plays a crucial role in the flavors produced during fermentation. Poor temperature control can have off-flavor implications that are not easily corrected. For post-fermentation treatments, our brewer has a whole new set of issues to consider.

Suppose our brewer selected a seven-barrel, three-vessel brewhouse constructed of stainless steel with steam heating and temperature control in the mash/lauter tun. The fermenters selected (to fit into the available space) were cylindro-conical, stainless steel, and jacketed vessels with a 2:1 height-to-diameter ratio and glycol cooling and large enough to ferment two brews with enough headspace so that they don't foam over. There is also glycol service for controlling fermentation temperatures to +1°C. Once constructed, the brewer now must take a 10-gallon recipe and scale it up to seven barrels (217 gallons). Some of this can be done with simple math, but much relies on trial and error to arrive at similar results in the glass—this is where the blend of art and science of brewing takes full shape. Starting with raw materials, since the ingredients for this beer were widely commercially available through well-known brewing industry suppliers, this was not an issue (in this case). There are, however, many concerns that have to do with how these raw materials are handled in the process. For example, the brewer has purchased a new grist mill that grinds the grain to a different fineness from the previous setup, so the extract efficiency achieved will be different. Temperature control in the mash allows for our brewer to hit desired temperatures much more efficiently, but chances are that there will be an impact on the levels of fermentable sugars that are produced. Lauter design will also impact extract efficiency. When we get to the brew kettle there will be impacts from the rate of temperature rise, extent of evaporation (this depends greatly on the surface area-to-volume ratio), hop extraction efficiency, and loss of hop volatiles. There could be color impacts as well. All these variables must be carefully evaluated as the brewer takes the recipe from one brewery to the next, and we haven't even made it to fermentation yet. Through multiple iterations of trial brews our brewer should be able to design a repeatable process that achieves the desired outcome every time.

Regardless of all the thought and planning that went into the first trial brew, the results are not likely to be a perfect flavor match, nor is it likely that the brewer will have hit all the desired specifications of alcohol, bitterness, color, etc. If the first brew had higher specific gravity than expected or more alcohol due to a more complete fermentation, then the adjustments are relatively simple for the brewer to hit the desired outcome and can likely be made rather easily by adjusting the grain bill or the mashing conversion time or temperature. The same can be said for bitterness—if it is low, add more hops.

An aroma issue can be a far more difficult problem to solve. If our pale ale has lost its estery profile in our new fermenters and developed a sulfidic off-note, there could be several factors at play. Our brewer will want to look at yeast pitching rate, aeration rate, possible contamination with wild yeast, the temperature profile of the fermentation, as well as other factors. In these instances, it is very tempting to change several factors at once, but that is not the most scientific approach. Our brewer may have read that esters are produced more favorably with a lower pitch rate, lower aeration, and higher temperatures; so, why not change all three? Our brewer might also have read that tall, skinny fermenters are not good for ester formation due to hydrostatic pressure on the yeast and therefore determine that the ester profile is no longer the most important factor for this beer. The brewer can then concentrate on removing the sulfidic note and focus on a set of trial brews with differing temperature profiles and better yeast handling procedures to address that issue. Trial and error and the varying of only one parameter at a time to arrive at the proper process to deliver a consistently delicious beer—this is the craft of the brewer, large or small.

Let us now imagine that our brewer successfully scaled up the beer to provide delicious beers to the thirsty brewpub clientele and that the brewpub was so successful that it was eventually purchased by a large brewing company. Aside from no longer being able to use the moniker *craft*, our brewer will have a whole new set of problems to face when it is time to bring this pale ale to a much larger drinking public and scale up to "industrial"-sized brews in a whole different scale of brewery. In this larger brewery, the brewer will face the exact same set of challenges, but they will be on a formidable scale. Trial and error with 700-barrel batches is a risky financial endeavor—even when it's somebody else's money. Once again, degree of milling, water-to-grain ratios, mashing profiles, and all the rest of the brewhouse parameters need to be compared and evaluated as the recipe is scaled up. The very same crafting of a recipe that happened when our brewer went from garage or kitchen homebrewer to professional brewpub brewer now needs to be done again but on a still larger scale. Once again, some of it is simple math, but some of it is much more complex, owing to the impact of physical factors on the complex microbiological, chemical, and biochemical processes that underpin the production of beer. The physical–chemical properties such as bitterness, color, and alcohol can likely be hit on the first or second trial, but the flavor match can be problematic indeed. One issue that our hypothetical brewer might discover is that the esters have come back when moving the fermentation to a fermenter that is more squat in shape. Now, that might seem like a good thing since this was part of the original lure of the award-winning homebrew, but now we are looking for a flavor match, and too many esters won't do—so, it's back to the investigation of yeast pitching rates, oxygenation, and temperature profiles. The craft is the same—a blend of art and science to arrive at the same goal, a delicious glass of beer that tastes just like the last glass of beer.

In the example of our pale ale, it is quite clear that size and scale do not matter all that much in the ultimate outcome of the beer. With modern brewing techniques and equipment, we can produce nearly identical beers in very different breweries and even with some differing raw materials. The scenario above has played out repeatedly in the modern brewing landscape. To see and taste the efforts of these brewers as they have scaled up, all one needs to do is visit the local grocery store. The examples include pale ales, India pale ales, diverse lagers, and many others. In all these cases size really does not matter. When might size be an issue? What specific styles only lend themselves to much smaller production volumes and cannot really be duplicated on a larger scale?

Some Things Are Never Big

There are several beer types and variations upon these styles that traditionally only lend themselves to relatively tiny batches due to the level of processing necessary. Many of them are being produced by larger brewing companies after they have acquired many smaller breweries and can use these facilities to produce these smaller-volume, "niche" beers. An example of this is barrel-aged products.

The beer must inherently be stored in individual barrels. This is a labor-intensive process that requires the acquisition of many used barrels from various sources, filling each of them with beer, and storing for an extended period before emptying these barrels back into a tank prior to packaging the beer. This sounds like the purview of much smaller breweries; however, Anheuser-Busch InBev (the largest brewer on the planet) has a large stake in this particular style through the Goose Island brewery (among others). Sour beer is another category where small-batch fermentation or post-fermentation treatments are traditionally performed in large barrels called "foeders," or various other vessels. These are now made by brewers large and small with varying degrees of success. Sour beers with spontaneous fermentations are a different sort of thing altogether and are usually only produced in smaller breweries; however, there is no real reason that they could not be brewed in larger brew sizes if the demand grew sufficiently. The larger processing footprint required for the coolships that are employed to cool the wort and seed the brew with adventitious microbes would necessitate some investment, but if the demand is there, they will be installed.

Beers with unique ingredients also lend themselves to smaller brew sizes, especially if the ingredients are hard to come by or incredibly expensive. The reasons for this are usually more commercial than process-related. There is no reason why we could not make a 700-barrel batch of a hibiscus and ginger–infused mild ale with saffron and rose hips added post-fermentation, other than the fact that we might have trouble selling all 700 barrels within the suggested shelf life of said beer. It also may be cost-prohibitive to acquire the necessary amounts of ginger, hibiscus, rose hips, and saffron to make a batch that large.

A Notable Example

One example of a beer that would be difficult to scale up much more than the level at which we currently make it is a beer that is near and dear to the heart of one of us, Anchor Steam Beer. To understand why, it is important to understand what a steam (aka California Common) beer is and how it is made. Technically, this is a beer fermented with a lager strain of yeast at a relatively warm temperature in an open, shallow-pan fermenter. It is a free-rise fermentation with no temperature control other than that afforded by the cool air in the room. This is a traditional style of product that has been brewed in San Francisco since gold rush times. The wort is cooled to 60°F (15.5°C) as it fills the fermenter, while the filtered air in the room is also cooled to a constant 60°F, but since the fermenters have no cooling jackets, the heat of fermentation causes the fermenter to eventually warm up to around 75°F (24°C) by day 2. It also foams up beautifully (Figure 1.1). By the morning of day 3 the yeast has dropped out to the trough in the bottom of the fermenter, the foam is gone, and the primary fermentation is done as the beer begins to cool back down toward room temperature. The beer is then dropped down to a cellar tank, where it is kräusened (addition of a vigorously fermenting new brew) with a 12-hour fermentation and allowed to have a forceful secondary fermentation in a closed vessel where it can naturally carbonate itself and gracefully age for at least 2 weeks at the relatively cool cellar temperature of 55°F (13°C).

The flavor implications of this fermentation approach are directly tied to the process and especially to the equipment used in this method. The wort is easy enough to replicate, but it is the fermentation that sets this beer apart. The fermenters are quite

Figure 1.1 A fermenter at the Anchor Brewing Company

shallow—only about 3 feet deep and with 15 × 30 feet of surface area. This has a huge impact on ester production—as does the warm fermentation. The result is a beer that has the crispness of a lager but also a floral and fruity, estery profile that is reminiscent of an ale. The kräusening process also has a great impact on the maturation of the beer and ensures a proper maturation. The complications of making a style like Anchor Steam Beer require engineering and designing the brewery specifically to make that style of beer—something that Fritz Maytag did in 1977 and nobody has done since. This is not to say that it cannot be done. If the desire was there to build a brewery with similar functionality and capabilities on a larger scale, it could be done with the right level of investment. This particular process is difficult to scale up to a larger brewery, but it is also quite difficult to scale down to a smaller process—our seven-barrel brewery example with pencil-shaped fermenters is an example of a brewery that should not attempt this style. Just one thing: the words *steam beer* are protected and can only be used by Anchor!

There are many other fine examples of historic breweries with processes designed over generations and that are intrinsic to the style of beer made. Many are in Belgium, but there are a few right here in the United States. The fine sour ales made by New Glarus Brewing in Wisconsin are examples of beers from a brewery built with a specific goal in mind. Russian River in northern California also has some fine sour beers made specifically in a beautiful coolship room designed for this purpose. The examples are many, and the common thread is brewers designing processes to meet their desired outcome. That upshot should always be measured one glass at a time, and hopefully that expectation is met, whether the brewery is large, automated, and awe-inspiring or tiny and completely manual.

References

Bamforth, C. W. Scientific Principles of Malting and Brewing. St. Paul, MN: American Society of Brewing Chemists, 2006.

Bamforth, C. In Praise of Beer. New York: Oxford University Press, 2020.

Brewer's Association. 26 (2005): 38–40. https://www.brewersassociat ion.org/statistics-and-data/craft-brewer-definition/, accessed January 6, 2023.

2

ARTIST WINEMAKER

ANDREW L. WATERHOUSE AND NICK E. GISLASON

Definition

As with other artisanal products, artisanal wine is presumed to be made in small quantities and with the care and attention of an artisan who has a vision of the product being created. However, due to the history and culture of winemaking, there are some important factors that don't come into play in other products. The first aspect to take into account is the location of the vineyard. Anyone with a passing knowledge of wine is aware that the geographic source of the grapes used to ferment the wine is key to the quality and reputation of the wine. When considering artisanal wine, and in particular notable wines in this category, the grapes most often come from a specific vineyard; and that vineyard is noted on the wine's label. US labeling further allows the use of the term *estate* when the grapes are specifically grown by the company making and selling the wine.

The wine produced at a particular site can also be radically different depending on the grapevine variety planted in that vineyard

Andrew L. Waterhouse and Nick E. Gislason, *Artist Winemaker* In: *The Science and Craft of Artisanal Food*. Edited by: Michael H. Tunick and Andrew L. Waterhouse, Oxford University Press.
© Oxford University Press 2023. DOI: 10.1093/oso/9780190936587.003.0003

(i.e., Chardonnay, Zinfandel, or Merlot). In the famous European winegrowing regions, there is no choice of what grapes to plant. There are regulations that require a specific variety or a short list that must be utilized if the producer plans to use the famous name of the region or "appellation." Well-known examples would include Burgundy (France), Chianti (Italy), Rioja (Spain), and Mosel (Germany). The regulations ensure that the wines that use those names have a typical taste, and the wrong grape variety will have a very different taste. In some regions, they even require that the wine pass a taste test for "typicity" (Saunders 2004).

Artisanal winemakers typically will describe their wines as being an "expression" of the vineyard or site where the grapevines grow. This is a complex issue since grapes do not turn into wine on their own, and many critical decisions must be made in cultivating the vines, harvesting the grapes, and then making the wine, each of which leaves a mark on the wine's final state. For instance, the grapevine's canopy (which can leave the berry bunches shaded or sunny depending on how it is pruned and hedged), the amount of irrigation, and the winter pruning techniques all have dramatic effects on the final composition of the grapes; and it is not possible to NOT decide. Even more profound effects are the picking date and crushing technique employed, and again, these mandatory decisions do not have default responses that could be described as "natural." Thus, the "expression" of any vineyard is in the context of a skilled winemaker's decisions which create wine with a style imprinted by those decisions. A major question we will explore is, can an artisanal winemaker's choices so influence the outcome that it might be possible to discern the artisan by tasting the wine?

Winemaking Vision

Any object or device can be made without particular regard for a thoughtful plan, for instance, trying to copy a popular painting or following a standard design for a table. However, when the plan involves consideration of history, cultural standards or expectations, the creative and expert use of technology, and creative changes in the plan, the outcome can be said to be driven by a vision. If that vision is applied consistently for a number of years, then that outcome could be defining a personal style, one that might be recognizable by cognoscenti.

Vineyard Site Selection

In the New World, where vineyards are, at most, 150 years old, and the vast majority much younger, the choice of grape variety is not encoded into law, at least not yet. So, aspiring artisanal winemakers have a liberty unheard of in the famous regions of Europe. They can seek out vineyard sites within established regions and plant any variety of *Vitis vinifera*, or other species or hybrids that they feel will produce a wonderful wine in line with their vision. In the better-known regions, such as Napa Valley, experimentation by growers over the previous 50 years or so has shown what varieties seem to do best in particular sites, or at least what sells for the highest price, so it may be costly to go against market expectations. But it is perfectly legal to grow Riesling in the Rutherford area of Napa Valley and call it Rutherford Napa Valley wine, even if 99% of your Rutherford neighbors are making and selling Cabernet Sauvignon (probably for 10 times the price of your Riesling). So, in the New World, the first theoretical step in crafting an artisanal wine vision is to select a grape variety, or a number of them, to plant in a specific site. In principle, an artist-winemaker has the liberty to choose any available site and plant any grape variety on that site, but planting within known regions (appellations) and, in the Old World, with allowed varieties provides a marketing advantage.

So, in the "New World" of wine, the first step in artisanal wine is to decide where to plant a vineyard and what grapes to plant there. Key factors in this decision will have to do with climate and soil, with water retention being the most important factor for soil characteristics (van Leeuwen et al., 2011). Certain temperature regimes (heat indexes) are well suited to particular grape varieties, so the choices are somewhat constrained. While market demand for particular wine grapes is certainly a factor, with five varieties currently comprising 90% of the market, hundreds of options are left to eke out tiny slices of the pie, though there is growing interest in these different tastes. In established Old-World vineyards under controlled appellations, these decisions are largely made for you, but there is still some flexibility in choosing which vineyard to utilize within a specific appellation, as well as a limited range of choices in varieties to plant and the specific clones for the required varieties.

Rootstock

Another issue in grape growing is the rootstock since most vine-yard regions have phylloxera in the soil, a pest that can kill V. vinifera. So, when planting V. vinifera, a grafted rootstock must be selected. While the primary factor is pest resistance, rootstocks present different levels of vigor and varying compatibility with particular soils (Jones et al., 2009). In California, the current practice is to select a rootstock that has as little vigor as possible for the site, so as to limit growth and enhance quality. However, a vigorous rootstock will be able to recover water from a greater volume of soil, and thus survive under difficult conditions with fewer resources, making the vineyard more sustainable.

Cultivation Practices

With a particular vineyard that has one or more grape varieties planted, the next step is to decide on how to cultivate (or not) the soil around the vines. While this has much less impact than the grape variety selected or the site, certain parameters can dramatically alter the outcome. For instance, the amount of water available to the vine has a major effect, so summer rain or irrigation choices will alter grape and wine quality. Dry farming in California, which essentially has no rain between May and October (i.e., no irrigation), could result in small vines with low crop yields, depending on the choice of rootstock. Some advocates claim this makes the best wine, but judicious irrigation can be used to make excellent wine. In general, lower yields are correlated with higher quality; but in a particular site, dropping crop (grape clusters) does not automatically lead to higher-quality product. Very low crop yields are a badge of honor in some corners, but the resulting wine may have to be sold at astronomic prices to cover costs, tarnishing sustainability credentials. The most authentic wine has been described as that wine made by artisans to reflect a site and their style (Kreglinger, 2016). As above, in controlled appellations, choices are generally constrained, with, for instance, irrigation forbidden in locales with usual summer rain.

Aside from irrigation, there are many practices that are quite important. Vines need pruning each winter, and this sets the stage for crop yield as well as canopy (vegetation) growth for the next vintage. Depending on leaf area, soil nutrition, and water, the vines will be able to ripen a particular amount of fruit, so getting this wrong can lead to grapes that don't ripen or have far too much canopy and suboptimal flavors. Soil health has many implications, in particular water retention and greenhouse gas absorption. In California, it is important to capture as much of the winter rainfall as possible, so a good groundcover of grasses will help the soil absorb rainfall, usually all of it, while bare soil will shed the water into streams and rivers, along with some soil, resulting in a greater need for irrigation (Faulkner, 1943). And the groundcover will provide live soils populated with a vigorous microbiome that can capture carbon dioxide from the air and help fix atmospheric nitrogen into plant-usable forms (Wagg et al., 2014).

True craft winemakers are making their product from a specific vineyard and thus need to have that source managed in a sustainable manner. While the perceptions of sustainable viticulture have evolved, current practices include reduced fertilization and pesticide use. For instance, it has been standard practice to spray elemental sulfur powder in the vineyard on a schedule to control molds, but by monitoring the temperature and humidity, it is possible to limit spraying to times when mold is expected to be a problem. Even better, it is possible to monitor for mold spores and spray only when an infection is imminent, greatly reducing the need for sulfur sprays (Rodriguez-Rajo et al., 2010). And today, the concept of sustainable winegrowing includes attention to soil health, by encouraging the presence and growth of soil microorganisms as well as supporting ecosystem services.

A major difference between artisanal winemaking and large-scale winemaking would be that the grapes in a particular wine will have come from more than one vineyard, in some cases from vineyards across a wide region, such as the state of California. And along with this style of sourcing, the grapes are being grown by a number of farmers, each with their own idea of how to cultivate their grapes. Purchase contracts can specify certain parameters; often, the criteria are the sugar level at harvest and crop yields, but other parameters are generally at the grower's discretion. This is one reason large-scale winemaking is not artisanal as the winemaker cannot have much input into how the grapes are grown, nor are they very involved in a critical decision, when to harvest.

In artisanal winemaking, each and every factor in cultivation is carefully considered with a goal in mind of a particular outcome in terms of grape composition and yield. The best wines are made, as many would say, "in the vineyard"; and artisanal winemakers actually focus more on the field than on the cellar to create a particular style. In fact, the CEO of a large, well-known wine company explained that the limit of his company's production was defined by the number of acres that his winemaker could monitor, in particular at harvest, for the next big decision (Pearson, 2015).

Harvest

Aside from variety selection and site, the harvest decision has the most impact on the character of the wine. As the grapes ripen, the amount of sugar increases; and this translates into the concentration of alcohol in the finished wine. It is possible and legal (in the United States) to add small amounts of water to decrease the sugar concentration at harvest in order to reduce alcohol or to add sugar (*chaptalization*) to increase alcohol, as practiced in many cool regions. However, and more importantly, the specific profiles of secondary natural products in the grape are changing each day during ripening. Some are increasing and some decreasing, and these profiles define the "varietal" flavor of the wine—in other words, the distinctive taste of the particular wine resulting after fermentation. As noted above, this decision is so important that the amount of truly artisanal wine made by any individual is limited by their ability to assess when to harvest the fruit.

While analyses are essential to following ripening, the harvest decision cannot be made by reading sugar and/or acid levels from laboratory measurements. In artisanal winemaking it is essential to walk the vineyard both to inspect the appearance of the grapes and vines and to taste the fruit, chewing the skins and/or seeds for their astringency and texture. Factors that are considered include the taste and texture of the skins and seeds but also the appearance of the skin on the grape, the color of the rachis, the tension holding the skin to the grape's flesh, and even the force needed to detach grapes from the pedicel. The skill of an artisanal winemaker has much to do with their ability to recall the grape flavor profiles they previously tasted and the resulting wine. This provides an ability to anticipate the wine outcome of the fruit under consideration. This is much akin to a painter anticipating how a newly applied paint will look after drying or a musician anticipating how music written on a score will sound in a performance.

Crushing and Fermenting

After harvesting (usually done at night or early morning for best quality), the grapes are brought to the winery. The first step is to sort through the clusters to remove leaves or stems and material other than grape from the vineyard. Then, high-quality reds are destemmed and the berries sorted in order to remove unripe or desiccated berries, selecting only the desired fruit for the anticipated wine style. Berry sorting is typically done by an automated sorter, but sorting crews on harvest lines are still common. For some red wines, destemming is avoided, and whole clusters are added to the fermenter. The amount or fraction of stem inclusion can vary from 0% to 100%, and the decision is an important stylistic one that has a significant impact on the aroma and taste of the final product.

In white winemaking, the fruit will be destemmed and pressed after sorting to yield juice. The juice is generally held cold overnight to settle out particles of skin and pulp that will influence wine quality. This semi-clarified juice is racked and then fermented, either in barrel, stainless steel tank, or concrete fermenter, depending on the desired style. The ferments are carried out cold, around 60°F, to retain fruitiness and will take 2–3 weeks. Ancient winemakers and a few today handle whites like reds, retaining the skins during fermentation, resulting in what is now called orange wines, which have much more tannin and body and are very different!

Depending on the climate, sugar (cool) or acid (warm) is added to balance the juice, though it is possible to make wine with no addition. Minimalists specifically pick grapes at a time when they anticipate that no addition will be needed for a balanced taste of acidity, alcohol, and sweetness. In general, if the grapes have been cultivated and harvested with care, then few or no adjustments are needed; and under such conditions, the wine is said to better reflect the qualities of the vineyard.

Rescue Artists

While minimal intervention is common among artisans, another approach is to craft a wine to a particular style with a winemaker's skill. Making adjustments is common in large-scale winemaking, where a particular style or flavor is desired, regardless of the grapes' origin. However, some well-known artisanal wines are crafted from varying grape supplies to achieve a particular flavor profile. The first step is to select a blend of grapes or wines to achieve the general flavor. The blends will be similar in most cases, but when a particular grape supply dries up, drastic changes may be needed. In red wines, some common blenders include Barbera for acidity, Zinfandel for fruity aromas, Petit Syrah for tannin, and dryer varieties such as Alicante Bouschet for color. After crushing, sugar can be added to boost alcohol, water to reduce sugar concentration, acid for tartness, and, later, oak (i.e., barrels) for sweetness, tannin for mouthfeel, or fining with protein to remove tannin. These are some common additives, and while others are permitted, oak is the only "flavoring" permitted. In addition, prior to bottling, other tweaks are common, such as a juice or juice concentrate addition for a touch of sweetness (this can have a dramatic effect even at levels that don't have any "sweet" taste) or low-level carbonation.

But other additives would be largely eschewed by artisans, while large-scale winemakers commonly add grape pigments to boost red color, tannin to adjust mouthfeel, and a tiny bit of grape juice (grape sugar) to the finished wine to add mouthfeel and reduce harshness. Some winemakers inoculate with yeast, while others allow ambient yeast in their facility to do the job. The latter approach means that fermentation times and outcomes are not as predictable, but this can be viewed as another aspect of the natural character of the site; and this practice is often marketed to consumers.

Red winemaking is more complex since the important color and tannin are in the skins and seeds and must be extracted during fermentation and afterward if more is desired. The ethanol produced during fermentation is the key extracting principle, and the temperature of the process has a very large impact. The procedures that facilitate extraction are called *maceration* by winemakers, and there are many options including the Burgundian tradition of *pigeage* (punch downs) or the classic pump-overs; but many other options exist. Each has the variations of interval, intensity, and length of process, all of which are altered throughout the fermentation and in response to the current state of the must. While a number of experiments have shown various impacts of these processes, artisans pay very close attention to the fermentations, tasting at least daily to assess the next process decision. In some wine styles, rather than pressing at dryness, the standard endpoint of maceration, if the pomace is left in contact with the wine for 2–20 days after (sometimes much longer), this extended maceration will increase tannin extraction, resulting in greater astringency and mouthfeel, attributes expected for long-aging red wines.

Chemists have shown that red grape tannin and pigments react during and after fermentation to produce a series of "wine" pigments that are derived from the grape pigments (anthocyanins) but are modified. This includes reactions with fermentation metabolites, such as acetaldehyde or pyruvic acid, but sometimes also involving phenolics like caffeic acid or condensed tannins. These reactions lead to thousands of pigments, in small to infinitesimal quantities, which collectively supply red wine with substantial color, generally a bit less than the original wine. To a chemist these are very interesting products that have good chemical stability and provide wine color for decades. However, the state of knowledge at present makes the information about the profile (if available) of limited value, so winemakers, artisanal or industrial, generally measure overall color parameters but do not utilize detailed information on the formation of these derived pigments.

It is thought that some of these transformations alter the sensation of tannin astringency, making it softer or smoother; but there is no well-developed understanding of how these different molecules would affect the sensory perception. With so many different pathways and products, it will be a while before that understanding is available.

Pressing

White grapes are pressed at crushing, and the fermentation occurs on the juice, generally at cool temperatures. However, for red wines, the grape solids are retained during fermentation and now have to be removed. The timing, of pressing as noted above, is a major decision and can occur any time from midway during sugar consumption to up to days or week later or, in rare cases, months afterward. In the instance of long delays, the process is not just extraction but transformation as well. The decision on when to press will be matrix (or wine)–dependent as the winemaker will be striving to achieve a balance of aromatic flavors and mouthfeel, among other factors, so in some vintages pressing may be quickened or delayed to attain the desired balance.

In addition to timing, the specific device is a major decision. The standard press is a large horizontal cylinder with a balloon inside. The solids are pressed against the sides of the tank while it is rotated, with the pressure and rotation varying according to secret protocols to achieve the desired level of tannin extraction and yield. In some facilities, a basket press will be used for red wine, a hydraulic plate pressing down on the pomace. This style yields more clear wine as the "cake" of pomace acts as its own filter.

As the pressing continues, more and more tannin is found in the wine, and every winemaker has a pressure point to stop collecting the wine and, if the pressing continues, to divert the output to a separate stream of "press wine." In some cases, small amounts of this may be added back for structure and body, but most will be sold off for blending or distillation.

Oxidation

Oxygen is important for yeast health as it is needed for certain cell membrane components, and it is often added by exposing a fermenting must to air or by directly adding air to the must. Oxygen additions during fermentation or afterward also allow the yeast to produce acetaldehyde, an important factor in color stabilization for red wines. Small amounts of oxygen during aging also reduce the vegetal character of some grapes such as Cabernet Sauvignon, eliminating the bell pepper aroma notes from pyrazines that arise in the grapes.

Excessive oxidation is detrimental, and white wines can be exceptionally sensitive. One study of Riesling found that exposure to just 3 mg/L of oxygen at bottling (1 mg/L is a typical target maximum during bottling) altered the taste of the bottled wine after 6 months of aging. Red wines generally need much more oxygen, such as a cumulative exposure to 50 mg/L of oxygen, in order to properly stabilize color. When excessive oxidation occurs, wines take on an aldehydic aroma that can be due to acetaldehyde but is more often a result of a number of other aldehydes caused by oxidation.

Aging

With the "finished" wine in hand, the next question is the aging regime. Traditional red wine aging utilizes oak barrels, and white usually utilizes limited barrels along with some tank aging. However, there are now a plethora of artistic choices that have interesting marketing angles as well. There are now multiple ceramic containers including concrete in various shapes, such as "eggs"; quivri tanks from Armenia or Georgia; different woods, such as chestnut or acacia; oak from unusual sources; and stainless steel containers of different sizes, with or without wood inserts. The various materials impart subtle or distinct flavor changes to the wine depending on the substance and the volume-to-surface ratio. In addition, some are semi-permeable and admit limited oxygen. The winemaker will use this aging period to reduce the heightened fruitiness from fermentation and to meld the aging flavors into the wine. While these effects have less impact overall than some other choices, they provide the winemaker with another artistic touch to the final product; and these choices are often consistent for particular wines. The final decision is on the closure, and the major choices are natural cork, technical cork, synthetic cork, or screwcap. In each case, there are now different variations available that affect the key performance criterion, the transmission of oxygen. Natural corks are traditional but can have the cork taint fault that will affect a small percentage of wine bottles, though vendors are offering testing to remove all tainted corks—for a price. Some technical corks have had the cork particles cleaned to remove all taint, while synthetics and screwcaps have no taint issues. For these latter three types, vendors offer closures with specified oxygen transfer rates, and that choice will have a clear effect on the development of wines in the bottle. So, winemakers will choose more or less oxygen exposure during aging in order to provide the desired flavors during the expected aging period.

Conclusion

In conclusion, can the thought or concern, the care that goes into crafting an artisanal product like wine, be perceived while tasting it? It is well documented that such appreciation is possible in the visual and auditory worlds, where those knowledgeable of those arts can attest that the beauty of an exceptional painting or a violin solo was created by a specific artist. Even those without an education in the field can at least suspect that a particular piece of art or music was created by a particular artist, such as Van Gogh or Bob Dylan. So, is it possible that winemakers can create wines that have such a distinctive signature that other experts can perceive their craft? Cannot an exceptional wine elicit a similar appreciation of distinctive beauty among those who appreciate fine wine? We have heard a story that one experienced taster accurately identified the winemaker and reported that he was so sure of this that he saw the visage of the winemaker in a glass of wine. We feel that is possible to detect the hand of a winemaker in a glass of wine. Despite vintage differences, the choice of grapes from a particular site that has consistent cultivation techniques, along with the impact of harvest timing, fermentation protocol, pressing decisions, and barrel practices, can leave a lasting signature on the wine. If not, how would it be possible for wine experts, such as masters of wine, to recognize wines from particular sources?

References

Faulkner, E. H. *Plowman's Folly*. Norman: University of Oklahoma Press, 1943.

Jones, T. H., B. R. Cullis, P. R. Clingeleffer, and E. H. Rühl. "Effects of Novel Hybrid and Traditional Rootstocks on Vigour and Yield Components of Shiraz Grapevines." *Australian Journal of Grape and Wine Research* 15 (2009): 284–292.

Kreglinger, G. H. *Spirituality of Wine*. Grand Rapids, MI: William B. Eerdmans, 2016.

Pearson, D. Personal Communication, 2015.

Rodriguez-Rajo, F. J., V. Jato, M. Fernandez-Gonzalez, and M. J. Aira. "The Use of Aerobiological Methods for Forecasting Botrytis Spore Concentrations in a Vineyard." *Grana* 49 (2010): 56–65.

Saunders, P. L. *Wine Label Language*. Buffalo, NY: Firefly Books, 2004.

van Leeuwen, C., P. Friant, X. Chone, O. Tregoat, S. Koundouras, and D. Dubourdieu. "Influence of Climate, Soil, and Cultivar on Terroir." *American Journal of Enology and Viticulture* 55 (2011): 207–217.

Wagg, C., S. F. Bender, F. Widmer, and M. G. A. Van Der Heijden. "Soil Biodiversity and Soil Community Composition Determine Ecosystem Multifunctionality." *Proceedings of the National Academy of Sciences of the United States of America* 111 (2014): 5266–5270.

3

ARTISANAL CHOCOLATE

MICHAEL H. TUNICK AND JERRY TOTH

Introduction

Chocolate is a semi-solid suspension of fine solid particles from sugar and cocoa (around 70% total) in a continuous fat phase consisting of cocoa butter (Afoakwa et al., 2007). The bean obtained from *Theobroma cacao*, known as the cocoa tree or cacao tree, is the source of cocoa solids. Forastero (Spanish: "stranger"), the most common cultivar of *T. cacao*, is native to the Amazon basin and is responsible for approximately 80% of the cocoa in the world. The other major cultivars are Criollo ("of local origin," from the Caribbean area) and Trinitario ("Trinidad," a hybrid of Forastero and Criollo developed in Trinidad (Beckett, 2008; Aprotosoaie et al., 2015). Criollo is the most prized and expensive of the common cultivars because it is considered less bitter and more aromatic and is more susceptible to disease than the other major cultivars. A number of varieties of *T. cacao* grow in tropical regions (Beckett, 2008). The rarest cultivar used for making chocolate is Nacional ("national"), an heirloom variety that is native to Ecuador.

Michael H. Tunick and Jerry Toth, *Artisanal Chocolate* In: *The Science and Craft of Artisanal Food*. Edited by: Michael H. Tunick and Andrew L. Waterhouse, Oxford University Press. © Oxford University Press 2023. DOI: 10.1093/oso/9780190936587.003.0004

Central and South Americans were the first people to cultivate *T. cacao*, brewing bitter drinks from the beans and often flavoring them with chili. Genetic research indicates that the place of origin was the Iquitos region of Ecuador and Peru (Motomayor et al., 2008). Spanish conquerors exported the beans to Europe in the 1500s, and the chocolate made from them eventually became popular as a sweet food (Atkinson et al., 2009). The protocol for producing chocolate developed after much experimentation. The processing steps for chocolate are different from those of any other food ingredient, allowing large-scale and small-scale manufacturers to work with the product in unique ways.

Artisanal chocolate is produced on a small scale and is noted for having a high cocoa content and a more extensive flavor profile. The *T. cacao* plants may be specifically selected for their superior genetics, and the cocoa beans are fermented with great care to ensure a high-quality product (Nascimento et al., 2020).

Harvesting

The trees produce brightly colored fruits, known as "cacao pods," which are about the size of a 24-ounce water thermos and oblong in shape. Inside the pod are 30–40 large seeds, each of which is encased in a sweet white pulp. These seeds are the source of chocolate. Chocolate makers refer to the seeds as "cacao beans" or "cocoa beans."

The first step in the process is to pull off a cacao pod from a tree. But, as preliminary as this step may seem, at this point the broad strokes of the flavor profile are already established. The genetics of the cacao tree itself, combined with the unique soil and climate conditions in which the tree has grown, have a significant influence on the aroma and flavor characteristics of the chocolate that is ultimately produced. In this way, genetics and terroir play just as important of a role with chocolate as they do with wine. They are not the only factors, however. From this point forward, all of the people involved in the process exert their own influences over the flavors that are perceived in the nose and on the palate of those who eventually taste the chocolate.

Once the pods are pulled off the trees, they are opened by hand, albeit with the help of a machete. The seeds are scooped out and compiled in large wooden boxes and covered with banana leaves, whereupon they are left to ferment for 3–7 days.

Fermentation

The fermentation process is one of the single most important steps in the entire chocolate-making process, particularly from a chemical standpoint. The length of fermentation depends on the cacao variety, ambient weather conditions, and the flavor objectives of the chocolate maker. During the fermentation process, the wet cacao is manually rotated at intervals of 24–48 hours. This serves two purposes: it helps evenly distribute the heat generated through fermentation, and it introduces oxygen into the wet mass. The first rotation is arguably the most formative because it triggers the transition from the anaerobic (lacking oxygen) phase of fermentation to the aerobic phase.

Yeasts ferment the glucose, fructose, and sucrose in the pulp during the first 2 days. The anaerobic reactions produce ethanol and a temperature increase to 40°C. Then, aeration from turning the beans causes bacteria to produce lactic and acetic acids, and the temperature within the fermentation box steadily rises to 45–50°C. The heat, alcohol, and acetic acid cause the germ within the bean to die and release enzymes, which is where the flavor architecture begins to take shape (Schwan and Wheals, 2004). In the final 3 days of fermentation, the colors characteristic of cocoa are generated from browning reactions involving polyphenols, proteins, and peptides. Fermentation also removes some of the tannins and acids that are naturally present. Tannins comprise 5%–15% of the bean's weight and bring astringent and bitter flavors to chocolate.

Drying and Sorting

Once fermentation is deemed complete, the cacao beans are moved onto drying racks. The most effective drying racks are constructed underneath transparent roofs, in structures that resemble greenhouses. Many people in Ecuador dry cacao beans on cement floors, and some people even do it on the side of the road. The main idea is to dry the cacao beans via sun and fresh air, but the pace of drying is an overlooked feature of the process. When the cacao beans leave the fermentation bin, they are still wet and gooey and smell like wine. If they are exposed to too much sun too quickly, the outer surface of the cacao bean will dry before the interior does. In this event, the residual acids from the fermentation process are trapped inside the bean, which can ultimately lead to an overly acidic bar of chocolate. Poor drying can be avoided by carefully controlling exposure to sun during the first few days of drying. Air flow is also important.

Once the moisture level of the cacao beans has been reduced to 7%, the beans are ready to load into jute sacks and be carried to wherever the bean-sorting process will take place. The beans are laid onto a stainless steel table and manually sorted one by one based on size, shape, and health. Malformed or undersized beans are painstakingly removed.

Roasting

The sorted beans are then roasted. The most common way to do this is with a drum roaster. The drum is mechanically rotated, which internally tosses and jostles the beans, while heat is applied to the surface of the drum, usually in the form of fire. The objective is to evenly distribute the heat, thereby uniformly roasting the beans. The importance of pre-sorting the beans according to size now comes into focus. The idea is to segregate beans of different sizes into different roasting batches. If all of the beans are in the same batch, either the small beans will be over-roasted or the large beans under-roasted.

The two main variables in roasting are temperature and time, which together represent what chocolate makers refer to as the "roasting curve." The processor typically roasts the beans for 5–120 min at 120–150°C, which drives off the volatile acids. During roasting, aldehydes and other products with chocolate flavor notes are produced from Maillard reactions and the Strecker degradation.

The roasting process is critical. In the production of single-origin dark chocolate, the goal is to highlight flavors and aromas that are unique to the particular harvest of cacao. Over-roasting the beans basically lays waste to the entire effort. Under-roasting the beans may cause the consumer to miss out on some of the desirable flavors that can be created during the roasting process, and the acidity may be unpalatable. An optimal roast is thus a delicate balancing act.

Winnowing and Grinding

After the beans are roasted, they pass through a machine that cracks the beans and winnows (air-separates) the husks. Nibs are cacao beans that have been roasted, cracked, and winnowed. The husks are separated from the nibs and used for mulch or to make chocolate tea.

Then comes the grinding process. Ball mill grinders are the most effective, although melangers (stone grinders) can also work. Grinding is usually performed at temperatures above the melting point of cacao, whereupon the cacao mass is liquefied. A good grinding machine will reduce the particle size to about 20–22 μm, which accounts for chocolate's silky-smooth texture when it ultimately melts in your mouth.

Once the chocolate is ground to a fine particle size, it is technically referred to as "chocolate liquor" or simply "cacao mass." In large-scale production, the cacao mass is mixed with sugar and/or other optional additives, such as extra cocoa butter, vanilla, and an emulsifier such as lecithin (sometimes in combination with polyglycerol polyricinoleate). A so-called two-ingredient bar, however, only includes cacao mass and sugar. A 100% chocolate bar may exclusively contain cacao mass, or it may also include extra cacao butter.

Conching

The mixture is then loaded into a conch machine or, in some cases, a melanger. Both machines effectively heat up the chocolate to a liquid state and then mechanically churn it for anywhere from a few hours to a few days. The cocoa mass is conched at 40–80°C for a few days, allowing the particles to be coated with cocoa butter and causing the final flavor of the chocolate to develop (Aprotosoaie et al., 2015; Schwan and Wheals, 2004). This process releases volatile acids from the chocolate and gives final polish to the flavor profile. Here again, a delicate balancing act is required of the chocolate maker. A heavily conched chocolate will taste flat but easily approachable. An under-conched chocolate will be more aromatic and complex but may be uncomfortably acidic. Similar to the roasting process, the two important variables are temperature and duration.

At this stage, or before roasting, mass-produced cocoa may be treated with alkali (a process developed in the Netherlands and known as "Dutching") to improve the color and flavor, to increase the dispersibility of cocoa powder in beverages, and to decrease astringency and bitterness. Cocoa made from superior beans does not have the high acid and bitterness of Forastero beans and, as such, does not need to be Dutched (Aprotosoaie et al., 2015).

Cocoa Butter

Some of the cocoa butter is pressed out of the cocoa mass to produce cocoa cake, which may be ground into a powder for various applications. The cocoa mass can be mixed with milk products (for milk chocolate), emulsifiers, and sweeteners. Production of dark chocolate does not include milk products.

The most critical raw material for chocolate is the cocoa butter. The fats and oils in cocoa butter are almost all in the form of triglycerides, consisting of approximately 27% palmitic acid, 34% stearic acid, 34% oleic acid, and 5% minor fatty acids (Beckett, 2008). Most fats in nature crystallize into three polymorphic forms, but the structure of cocoa butter leads to six forms with unique solidifying and liquefying properties. Producing the correct crystal form for obtaining the proper texture and melting characteristics requires subjecting the chocolate to specific heating and cooling cycles, known as "tempering." In producing milk chocolate, mixing the milk fat and the cocoa butter decreases the melting points of both fats, requiring changes in the tempering steps (Afoakwa et al., 2007).

Labeling Requirements

Regulations in the United States require milk chocolate to contain at least 10% chocolate liquor and 12% whole milk (usually in dried form). Bars of fine milk chocolate often contain 30%–45% cacao, whereas cheaper chocolate can have as little as 5% cacao. Dark chocolate, which has no milk, contains 15%–35% chocolate liquor with cocoa butter, vanilla, sweetener, and usually lecithin as an emulsifier. White chocolate, which has no cocoa mass or cocoa liquor, is composed of at least 20% cocoa butter along with sugar, milk solids, and optional flavorings such as vanilla (US Food and

Table 3.1 Typical Composition (grams per 100 grams) of Chocolate Types

Component	Dark	Milk	White
Protein	7	7	7
Saturated fat	24	20	19
Unsaturated fat	19	11	12
Sugar	29	56	60
Other carbohydrates	19	4	0

Source: US Department of Agriculture, Agricultural Research Service, Nutrient Data Laboratory (2020).

Drug Administration, 2020). Table 3.1 shows a comparison of the composition of dark, milk, and white chocolates.

Some chocolates are not consumed directly:

- Baking chocolate (also called "bitter chocolate" and "unsweetened chocolate"), which is made from pure chocolate liquor (100% cacao with no sugar added)
- Bittersweet chocolate, which is dark chocolate with sugar and cocoa butter added and at least 35% chocolate liquor (70%–100% cacao)
- Semi-sweet chocolate, which is dark, sweetened chocolate made with at least 15% chocolate liquor (McGee, 2004, 645–712).

Sensory Attributes

People derive pleasure from eating chocolate by the ways in which it affects all five senses: sight, feel, sound, aroma, and

Table 3.2 Compounds Responsible for Sensory Attributes in Chocolate

Class	Specific compounds	Attributes
Acids	Acetic acid	Vinegar aroma
	3-Methylbutyric acid	Sweaty aroma
Alcohols	2-Phenylethanol	Honey/floral aroma/flavor
Aldehydes	2-Phenyl acetaldehyde	Honey/floral aroma/flavor
	2-Methylpropanal, 3-methylbutanal	Chocolate aroma/flavor
Esters	Various (mostly methyl esters)	Fruity aroma/flavor
Fatty acids	Palmitic, stearic, oleic acids	Creamy mouthfeel
Furanones	Furaneol	Caramel flavor
Pyrazines	2,3,5-Trimethylpyrazine	Cocoa/nutty aroma
	2,3,5,6-Tetramethylpyrazine	Cocoa/chocolate aroma
Pyrones	Maltol	Caramel flavor
Pyrroles	2-Acetyl-1-pyrrole	Caramel/chocolate flavor
Various	Various products of enzymatic and nonenzymatic browning reactions	Brown appearance

Source: Tunick and Nasser (2020).

taste. Table 3.2 lists key compounds responsible for sight, aroma, and taste.

A high-quality piece of chocolate should appear smooth and shiny with a color ranging from mahogany to black. When broken off of a bar, a clear and crisp snap should be heard. If the piece has a high cacao content and is well tempered, a clean snap should be experienced. If it splinters upon breaking, the chocolate is too dry; and if it resists breaking, it is too waxy. A crumbly break is also undesirable (Beckett, 2008). The snap will not be as pronounced with milk chocolate, which has lower levels of cocoa solids, or white chocolate, which has no cocoa solids. A piece should quickly start to melt in the hand with no graininess in the mouth, so the tactile sense is very important. The unique properties of cocoa butter are responsible for the mouthfeel and textural properties of chocolate (Hoskin, 1994). The amount of cocoa butter and the addition of surface-active ingredients (especially lecithin) will control the viscosity. Consumers perceive food to be gritty or coarse in the mouth if the maximum particle size is above 30 μm, so chocolate is milled to a maximum particle size of 22 μm, which imparts a creamier taste and texture (Afoakwa et al., 2007).

The aroma of chocolate may have hints of fruits, nuts, spices, flowers, or sugar. The most important odor-active compounds in chocolate are pyrazines; around 80 of these compounds contribute to overall flavor. Pyrazines generally originate from α-aminoketones through the Strecker degradation and Maillard reactions that occur during roasting. Esters, which arise from amino acid degradation and fermentation, are the second most important components contributing to characteristic aromas. Fermentation and amino acids also generate alcohols, aldehydes, and ketones. Acids include acetic acid (the most odor-active) and various fatty acids. Processing mostly removes short chain acids since these lead to undesirable odors. Other compounds include furanones and pyrones, which are produced by degradation of monosaccharides during drying and roasting, confer pleasant caramel notes, and enhance flavor impression (McGee, 2004).

The basic taste of chocolate should include sweet, bitter, and slight notes of acid, sour, and salt. The processing variables and inherent characteristics of the cocoa bean determine the taste. More than 600 volatile compounds, some of which contribute to flavor, have been detected in chocolate (Ziegleder, 2009). Flavor precursors develop during fermentation and primarily interact at roasting temperatures. Complex browning reactions also occur during roasting. The numerous heterocyclic flavor compounds produced then contribute to the characteristic and complex chocolate flavor (Aprotosoaie et al., 2015).

About 80% of the cells in cacao beans contain protein and cocoa butter for the nourishment of the plant, and the remainder contain defensive compounds meant to ward off animals and microbes. These deterrents include bitter alkaloids such as methylxanthines (including caffeine and theobromine), astringent phenolic compounds, and anthocyanin pigments (Owusu et al., 2020). Artisanal chocolate containing a high level of cocoa is a good source of caffeine, catechin, and epicatechin (Nascimento et al., 2020).

Spotlight on To'ak Chocolate

To'ak Chocolate is an artisanal manufacturer in Manabí province, Ecuador. The company, whose name is a combination of *earth* and *tree* in local dialects, uses Nacional beans grown in the valley of Piedra de Plata. These trees are located on hillsides and are often cultivated in combination with other fruit trees. Certain heirloom varieties in particular grow well in the shade of larger trees, hence their forest-friendly reputation among conservationists. The cacao trees are not irrigated, resulting in differences in the product from year to year due to weather. The terroir is also affected by the Pacific Ocean 70 km to the west of the growing area and the Andes Mountains located 70 km to the east.

During the first few days of drying the beans, their exposure to the sun is controlled with a wooden rake-like implement and, in many cases, with bare hands. At 4-hour intervals throughout the day, the workers run their hands through all the beans on the drying rack, akin to scrambling eggs on a frying pan so that no portion of the eggs is overly exposed to the heat of the stove. This is also an opportunity to physically connect with the cacao beans. Drying cacao is the most intimate step in the entire process.

To'ak has found that the best use of rejected beans from the sorting process is simply to eat them, raw and straight out of hand. The bean-sorting process can be long and tedious; therefore, frequent doses of theobromine, a compound similar to caffeine, serve to sharpen the senses and strengthen one's resolve to continue.

To'ak's bars are hand-crafted with a single roasted Nacional cacao bean selected individually and placed in the center to remind the consumer that the product is derived from the fruit of a tree. The bars are boxed in locally grown wood and come with tongs to prevent skin oils from coming in contact with the product. The company is studying long-term aging of its dark chocolate to determine its effects on flavor, texture, and aroma. To'ak's stated goal is to create for the world something utterly unique and beautiful, something that transcends merely being a "thing" and enters into the realm of experience (Toth, 2022).

References

Afoakwa, E. O., A. Paterson, and M. Fowler. "Factors Influencing Rheological and Textural Qualities in Chocolate: A Review." *Trends in Food Science & Technology* 18 (2007): 290–298.

Aprotosoaie, A. C., S. V. Luca, and A. Miron. "Flavor Chemistry of Cocoa and Cocoa Products: An Overview." *Comprehensive Reviews in Food Science and Food Safety* 1 (2015): 73–91.

Atkinson, A., M. Banks, C. France, and C. McFadden. *The Chocolate and Coffee Bible.* London: Hermes House, 2009.

Beckett, S. T. *The Science of Chocolate.* Cambridge: RSC Publishing, 2008.

Hoskin, J. C. "Sensory Properties of Chocolate and Their Development." *American Journal of Clinical Nutrition* 60 (1994): 1068S–1070S.

McGee, H. *On Food and Cooking.* New York: Scribner, 2004.

Motamayor, J. C., P. Lachenaud, J. W. da Silva e Mota, R. Loor, D. N. Kuhn, J. S. Brown, and R. J. Schnell. "Geographic and Genetic Population Differentiation of the Amazonian Chocolate Tree (*Theobroma cacao* L)." *PLoS One* 3, no. 10 (2008): e3311.

Nascimento, M. M., H. M. Santos, J. P. Coutinho, I. P. Lôbo, A. L. S. da Silva, Jr., A. G. Santos, and R. M. de Jesus. "Optimization of Chromatographic Separation and Classification of Artisanal and Fine Chocolate Based on Its Bioactive Compound Content Through Multivariate Statistical Techniques." *Microchemical Journal* 152 (2020): 104342.

Owusu, M., M. A. Petersen, and H. Heimdal. "Effect of Fermentation Method, Roasting and Conching Conditions on the Aroma Volatiles of Dark Chocolate." *Journal of Food Processing and Preservation* 36 (2012): 446–456.

Schwan, R., and A. Wheals. "The Microbiology of Cocoa Fermentation and Its Role in Chocolate Quality." *Critical Reviews in Food Science and Nutrition* 44 (2004): 205–221.

Toth, J. "Why the Most Expensive Chocolate in the World?" To'ak, July 17, 2022. https://toakchocolate.com/blogs/news/why-the-most-expensive-chocolate-in-the-world.

Tunick, M. H., and J. A. Nasser. "The Chemistry of Chocolate and Pleasure." In *Sex, Smoke, and Spirits: The Role of Chemistry*, edited by B. Guthrie, J. D. Beauchamp, A. Buettner, S. Toth, and M. C. Qian, 33–41. Washington, DC: ACS Books, 2020.

US Department of Agriculture, Agricultural Research Service, Nutrient Data Laboratory. "USDA National Nutrient Database for Standard Reference." 2020. https://www.ars.usda.gov/ARSUserFiles/80400535/Data/SR/SR28/reports/sr28fg19.pdf.

US Food and Drug Administration. "Code of Federal Regulations, 21CFR163." 2020. https://www.ecfr.gov/current/title-21/chapter-I/subchapter-B/part-163.

Ziegleder, G. "Flavour Development in Cocoa and Chocolate." In *Industrial Chocolate Manufacture and Use*, 4th ed., edited by S. T. Beckett, 169–191. Oxford: Blackwell Publishing, 2009.

4

ARTISANAL COFFEE

KELLY SANCHEZ, AUSTIN M. MROZ, AND CHRISTOPHER H. HENDON

Where do Coffee Qualities Come From?

For a foodstuff that is so ubiquitous, the line between artisanal and conventional coffee is a blurry one. As a consumer, the act of obtaining either is essentially the same. So why should one spend $6 on a cup of 93-point, natural-processed, delicately prepared heirloom-varietal coffee from Yemen and have a barista recommend you enjoy it black at 62°C? What makes it unique from the cheaper offerings at a local supermarket, perhaps paired with sugar and creamer, all closer to $1? To discover their differences, one may be inclined to use an analogy to another artisanal beverage such as wine. But this would miss a unique aspect of coffee—point-of-consumption preparation. Beyond the brewing, however, there are numerous physical and chemical considerations that govern the price and quality of coffee. To appreciate coffee's potential as an artisanal beverage we must first understand the journey coffee beans take, fancy and frugal alike, before being subjected to water

Kelly Sanchez, Austin M. Mroz, and Christopher H. Hendon, *Artisanal Coffee* In: *The Science and Craft of Artisanal Food*. Edited by: Michael H. Tunick and Andrew L. Waterhouse, Oxford University Press.
© Oxford University Press 2023. DOI: 10.1093/oso/9780190936587.003.0005

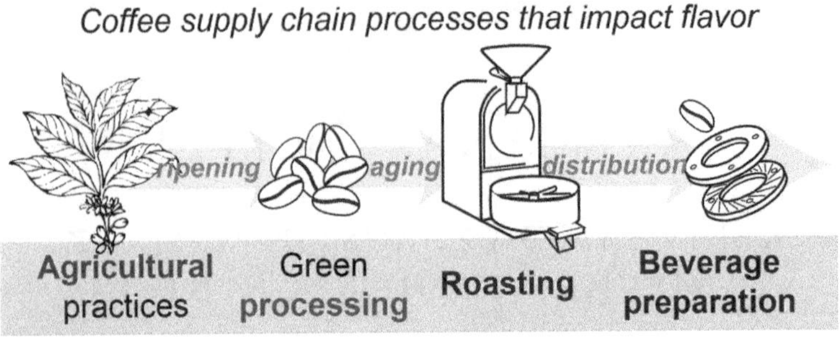

Coffee supply chain processes that impact flavor

Figure 4.1 Several variables determine the quality of a cup of coffee. Here, we present a simplified version of the supply chain, emphasizing the processes that are thought to make a major impact on terminal qualities of coffee.

to yield liquid coffee (Figure 4.1). The goal of this chapter is not to address the gulf between the commodity and artisanal camps but rather to acknowledge the elegance in both sides and ultimately explore the chemical and physical aspects of these extremes and preparations in between.

Coffee: Arabica and Robusta

Two primary varieties of coffee plants of the subgenus *Coffea* are prominently used in the coffee industry. These plants, *Coffea arabica* and *Coffea robusta*, yield "Arabica" and "Robusta," respectively (Lashermes et al., 1997), and account for approximately 99% of world production of coffee-identifying beverages (DaMatta and Cochicho Ramalho, 2006). Arabica coffee beans are generally considered to be more valuable than Robusta as they tend to produce more desirable beverage flavor (a discussion of this grading system is included Flavor and Aroma Descriptions). Both species also produce a wealth of generally negative flavors, ranging from astringency (e.g., chlorogenic acids [Bicchi et al., 1995]) to bitterness (e.g., caffeine [Dsamou et al., 2012]), in addition to crop defects. More broadly, variations in concentration of flavor precursors have been found to be considerably larger in Robusta than Arabica (Alonso-Salces et al., 2009; Fox et al., 2013). This may be due to the Arabica species possessing both low genetic diversity and low tolerance for environmental challenges, requiring more specific and uniform growing conditions. In general, Robusta is more resilient to a suboptimal environment and is higher-yielding than Arabica, though marked by a more variable and less favorable flavor profile (DaMatta and Cochicho Ramalho, 2006). Yet, Arabica has been shown to produce the most highly desirable flavor precursor composition at high altitudes, where CO_2 levels are sufficiently low to promote seed maturation. The emphasis on the dissimilarities between these two species is likely to only be intensified with changes in climate, with projections suggesting that over 60% of coffee species will soon face extinction (Stévart et al., 2019). For the most part, however, artisanal beverages are generally composed of Arabica coffees from either a single producing region or a blend of producing regions. With only a few exceptions, the artisans in the coffee industry use exclusively Arabica, and hence the remainder of this chapter will focus on the preparation of Arabica-based coffee beverages.

Soil Considerations

At their core, coffee quality and expression begin at the soil. Yadessa and colleagues have shown that, for Ethiopian coffees, calcium- and phosphorus-rich soil leads to coffee beans with increased size and cup quality, respectively. Increased nitrogen and sand content (slight acidification due to acidic sites on the surface of silica) tends to yield beans with lower caffeine levels (Yadessa et al., 2019, 2020) Mintesnot and colleagues (2015) further studied the role of early transition metals on coffee quality, indicating that the presence of iron was favorable for terminal flavor, whereas zinc and copper led to reduced enjoyment in the final product. Rekik and colleagues (2019) arrived at a more general conclusion, albeit for a more specific Colombian producing region, indicating that soil deficient in essentially all of the aforementioned minerals yielded improved quality. These studies, together with the industrial truth that coffee cherries are sorted based on size (see, e.g., Kenyan coffee grading systems—AA [largest], AB [smaller]), lead one to conclude that the negative impact of certain minerals outweighs the positive impact of others. It also points to a present necessity for grading based on flavor assessment as there are clearly other variables that govern coffee quality beyond mineral composition in soil and other a priori assessments.

Growing Considerations

Coffea is a flowering shrub in the family Rubiaceae. It bears fruit, or drupes—coffee "cherries" or "berries"—which, when ripe, are reminiscent of cranberries. Under-ripe coffee cherries have been shown to contribute heavily toward an astringent and bitter flavor profile and, in some cases, the corollary (heavily acidic, although roasting plays a significant role in the expression of these flavors [Dias et al., 2012]), likely due to the increased furan and hydrocarbon concentration. Indeed, flavor precursor composition significantly changes over the course of coffee bean growth (Ortiz et al., 2004) (Figure 4.2). Volatile composition of over-ripe cherries

ARTISANAL COFFEE

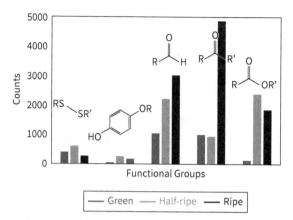

Figure 4.2 The chemical profile of coffee beans changes with degree of ripeness, encouraging selective harvesting to access desirable flavors. Ripe cherries exhibit very high levels of ketones compared to earlier harvested analogues.
Figure generated using data presented by Ortiz et al. (2004).

is not included in Figure 4.2 because the concentrations are so high and would skew the results (De Moraes and Luchese, 2003). However, it is rather the seed, not the fruit, that is highly sought after; hence, targeting ripe fruit generally leads to higher-quality coffee products.

It takes about 1 year for coffee fruit to develop its seed and fully ripen (Illy and Viani, 2005). With the exception of Brazil, which has mechanized the production of coffee, most producers cultivate and harvest coffee by hand. A knowledgeable farmer may invite a chorus of biodiverse plants and wildlife into the farm, but the best stewards on a farm are simply clearing the way for unique regional conditions to fatten the seeds with sugars and exotic acidities. A seed that didn't get this custom service will still undoubtedly taste of familiar "coffee" post-roast—it is the pampered plants that taste *different* and stand apart.

Processing

Post-harvest, coffee is sorted in a variety of ways: by size, degree of ripening, and blemishes. This can be done mechanically, with computer aid, by hand, or in some cases by floating the cherries and isolating beans based on effective density. For obvious reasons, the main objective of harvesting the coffee cherries is removal of the coffee bean from within; this step is termed *processing* and encompasses a variety of methods. It should also be noted that coffee processing is one area of active scientific development, with several unanswered questions. We will do our best to summarize the processes.

The color of a coffee cherry can vary depending on variety (e.g., Yellow Catuai, as the name suggests, ripens to yellow), but the general red-ripening form is common among Arabica. There is a fruit skin that encompasses the fruit mucilage, which in turn surrounds a seed(s) wrapped in a protective layer of parchment. Sometime during processing, all coffee is fermented (much the same as cacao, see Chapter 3 for more on this). In coffee, the mucilaginous layer surrounding the beans contains potential flavor precursors such as high concentrations of pectins (including galacturonic acids that had undergone both methylation and acetylation) and neutral polysaccharides (including monosaccharides such as arabinose, xylose, and galactose) (Avallone et al., 2000, 2001a, 2001b). During the fermentation process, these compounds become degraded and metabolized by indigenous microflora, potentially contributing positively to the flavor profile of wet-processed coffee, in particular (Lee et al., 2015; Zhang et al., 2019; Prakash et al., 2022). Due to variability in population of microflora, however, including bacteria, yeasts, and fungi, underfermentation and overfermentation are common and marked by unregulated growth of undesirable organism populations.

The *washed* process may be employed, where the skin and part of the fruit mucilage are removed from the coffee beans by wet depulping mechanisms, leaving the inner mucilage and seed (and protective layers) intact (De Moraes and Luchese, 2003). The sticky seed then undergoes wet fermentation, in which the inner mucilage is loosened and becomes separated from the coffee bean. The resultant green coffee bean is then sun-dried inside the remaining parchment layer, which is removed during a subsequent dehusking process. Wet processing has remained a popular practice as a means of obtaining desirable citrus, floral, and caramelized flavors, especially when applied to higher-quality and Arabica coffees (Zhang et al., 2019). It is widely believed, but unconfirmed, that fermentation can enhance the perception of desirable acidity in the roasted coffee. In American retail coffee culture the explanation behind this phenomenon is more fable than chemical as the proof is basically qualitative in that wet-processed coffee tastes "brighter" or more acidic than coffee processed in other styles.

Dry processing, also known as *natural* processing, is common in arid regions such as Ethiopia, Yemen, Brazil, and Panama, where harvested coffee cherries are sun-dried in large batches. Controlled drying and periodic turning of the cherries result in a product similar to a raisin. The process relies on agitation and ventilation to prevent fermentation and molds, fungi, insects, and bacteria from contaminating the crop. Coffee beans dried this way will often vary significantly in moisture content and are sometimes mechanically dried instead. Uncontrolled and extremely high temperatures using this method can result in "stinker" beans (i.e., coffee with fermentation defects and other unpleasant odors and tastes) (Vincent, 1987). The dried pulp of the coffee cherry is then detached using mechanical friction, and the seed is removed from the remaining layers in a dehusking process. In general, coffee beans that undergo dry processing are characterized by a notable fermented fruit flavor that, when darkly roasted, can obtain a pungent and even skunky smell.

And somewhere between the two aforementioned lies the *pulped* process (semi-dry), a technique that has gained popularity, particularly in Brazil (Duarte et al., 2010). Here, the fruit pulp is removed, and the coffee beans are dried while still inside an inner mucilage (Vilela et al., 2010). The dried product is finally dehusked much the same as a washed coffee. Instead of using water to remove the inner mucilage, however, the resultant beans are then dried with the remaining inner mucilage intact, which generally produces an intermediate body between the wet and dry methods and has seen good success in coffees developed for espresso. Like the natural process, the semi-dry process promotes flavor generation from consumption of the inner mucilage by microflora but is understood to omit sour and alcoholic flavor defects common to the natural process and overfermentation. Additionally, due to the mucilage having largely been removed, semi-dry processing has been shown to produce decreased concentrations of trigonelline and chlorogenic acid, at times responsible for a negatively astringent and bitter flavor (Barbosa et al., 2019).

Further, there are a handful of less common processing methods that are regionally specific: Kopi Luwak in Indonesia, where a civet digests and ferments the coffee; Giling Basah in Sumatra, where the beans follow a similar protocol to washed processing but are removed from their parchment while still wet; and Monsooned Malabar in southwest India, where the coffee is exposed to monsoon rain, resulting in decreased acidity (Davids, 2001). There are also many other "designer" processing methods (e.g., fermentation in the presence of pine needles, cinnamon, and so forth [Gu, 2017]), as well as anaerobic (Avallone et al. 2001a, 2001b) and carbonic macerated processing (Drioschi et al., 2021), the latter being borrowed from wine fermentation (Hickinbotham, 1986), highlighting the significance of the fermentation step in coffee flavor production. In all cases, overfermentation is a primary source of defective and off-tasting coffee (Lee et al., 2015).

Independent of the processing method, the coffee is then bagged and exported and spends significant time (~3 months) proceeding to the green coffee importer facility in the consuming country. While this process undoubtedly affects the coffee, there are no rigorous scientific studies on the degradation of the coffee because the storage conditions and length of storage time are highly variable. Once the coffee arrives, however, it will eventually make its way to a roastery, where the roaster will turn the green beans brown.

Roasting

Fundamentally, roasting is the transformation of turning green coffee brown, while developing both flavor and aroma (Buffo and Cardello-Freire, 2004). During roasting, green coffee beans are heated, initiating several thermally activated chemical reactions (Toledo et al., 2016). This process is thought to yield both flavor compounds and chromophores (making coffee brown). Practically, the general browning process is thought to occur above ~120°C, and some steps become more rapid at temperatures exceeding 175°C. Besides making and breaking organic bonds, the process also releases CO_2 and H_2O from within the bean, resulting in audible and physical cracks. All coffees are roasted to at least first crack (see Figure 4.3).

Many of the reactions initiated during roasting are effectively irreversible (Rivera et al., 2011; Iaccheri et al., 2019). Both the physical and chemical changes may be monitored using standard analytical techniques, including nuclear magnetic resonance (NMR) spectroscopy (Wei et al., 2012) and ion mobility spectrometry-mass

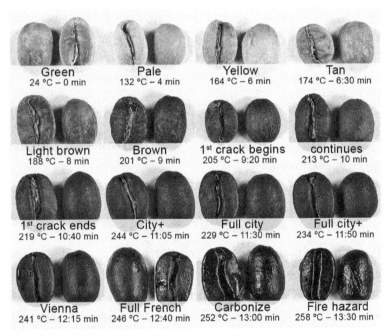

Green	Pale	Yellow	Tan
24 °C – 0 min	132 °C – 4 min	164 °C – 6 min	174 °C – 6:30 min
Light brown	Brown	1st crack begins	continues
188 °C – 8 min	201 °C – 9 min	205 °C – 9:20 min	213 °C – 10 min
1st crack ends	City+	Full city	Full city+
219 °C – 10:40 min	244 °C – 11:05 min	229 °C – 11:30 min	234 °C – 11:50 min
Vienna	Full French	Carbonize	Fire hazard
241 °C – 12:15 min	246 °C – 12:40 min	252 °C – 13:00 min	258 °C – 13:30 min

Figure 4.3 While each roast profile (i.e., temperature and time) is different, the general coffee roast begins with a "green" bean that becomes progressively browner, before audibly cracking. After the first crack, the coffee may crack again prior to the bean evolving oils (Vienna and beyond), eventually becoming fully carbonized.

Images produced with permission from Sweet Maria's Home Coffee Roasting.

spectrometry (Gloess et al., 2018). Additionally, theoretical models for heat and mass transfer have been developed (Fabbri et al., 2011; Fadai et al., 2017). The latter increases the internal bean pressure and temperature to above its glass transition temperature (T_g), as well as an overall volume expansion (we note that Figure 4.3 does not include scale bars, and hence does not properly convey the true volume expansion—in reality, the beans nearly double their volume upon roasting [Iaccheri et al., 2019]). Glass transitions are a phase change in an amorphous material between the solid- and liquid-like states. Below T_g, molecules are frozen within the amorphous structure and only exhibit rotational/vibrational motion. Above T_g, translational movement is afforded, and the mixture acts like a supercooled liquid (Roos, 2010). Significantly different material properties are associated with chemical systems at either end of this temperature spectrum, which is exacerbated by the roasting process.

One study sought to elucidate the specific mechanism behind the motion of localized molecules in glassy structures within roasted coffee beans using a combination of measured calorimetric and dielectric properties (Iaccheri et al., 2019). There, the authors posited that both green and roasted coffee beans are composed of locally ordered matrices. This result is somewhat surprising as it infers that, despite being crystallographically amorphous, there is some repeating order within the coffee. T_g was measured using differential scanning calorimetry, which revealed an indirect correlation between T_g and water/glycerol content. T_g was determined to be 34.89 ± 0.02°C to 48.76 ± 0.04°C for green coffee beans and 40.15 ± 0.1°C to 45.73 ± 0.05°C for roasted coffee. These are remarkably low T_g values, indicating that coffee could reach its glass transition in mildly elevated temperatures.

Changes in the lipid profile during the roasting process may be observed by the formation of aldehydes as a result of the degradation of hydroperoxides and other similar compounds, which has been monitored using ^{1}H-NMR (Williamson and Hatzakis, 2019). In the same study, roasting had an overall acceleratory effect on the hydrolysis of triglycerides, based on the increased free fatty acid concentration. Indeed, extraction of coffee oil from roasted coffee beans reveals that linoleic acid and palmitic acid are the main contributors to the total fatty acid content, followed closely by stearic and oleic acids (Calligaris et al., 2009). Notably, linolenic and arachidonic acids account for about 1%–2%, while gadoleic and behenic acids are only found in traces. Linoleic acid concentrations, specifically, are of note because humans can only obtain these through consumption (Dong et al., 2015); and one study showed a decreased likelihood of developing Type II diabetes (Lankinen et al., 2019). Further, it should be noted that most of these fatty organic compounds have poor solubility in water— their presence is primarily diagnostic, or a secondary indication of flavors in brewed beverages. There was no significant difference in the saturated fatty acid content measured before and after roasting (Williamson and Hatzakis, 2019).

In addition to the aforementioned non-polar compounds, diterpenoid alkaloids are commonly found in green coffee. Upon roasting, diterpenes degrade to form dehydrated derivatives (Dias et al., 2014). While the scientific community seems to agree that the total amount of accessible lipids increases over the course of roasting, studies disagree about the kahweol and cafestol concentrations: some report that their concentration remains unchanged, while others argue that it decreases with roasting (Williamson and Hatzakis, 2019). Stable levels of cafestol and kahweol upon roasting are counterintuitive because of the observed formation of their dehydrated derivatives, although stability may be due to the reformation of kahweol and cafestol by a yet unknown reaction pathway. Further, polysaccharides account for almost half of the chemical composition of green coffee beans; however, they do not appear in analyses of green coffee extract since they compose the insoluble polysaccharide complex of the cell wall (Wei et al., 2012). Upon roasting, the cell wall expands and ruptures and the polysaccharides depolymerize, thereby increasing the solubility of arabinogalactans and mannans. These water-soluble polysaccharides contribute to the chemical profile of roasted coffee extract; improve the retention of volatile molecules, known to contribute to the coffee aroma; and increase the perceived viscosity of the brewed extract (although the experimental viscosity may not have been experimentally measured).

Perhaps the most notable chemical transformation is the Maillard reaction (van Boekel, 2001, 2006; Ames 1996), a complex multistep cascade between alcohols, carbonyls, and amines (generally thought to be a reaction between sugar-like products with biologically relevant amino acids, see Figure 4.4). The process is thought to generate low–molecular weight acids (Scholz and Maier, 1943) and countless volatile and non-volatile organics (Figure 4.4) (Lund and Ray, 2017). While the exact mechanism of this non-enzymatic browning is complex, the reaction is described by three general stages: i) an initial condensation between a sugar and an amino acid, ii) formation of sugar fragmentation products via release of an amino group, and iii) formation of flavor compounds by subsequent fragmentation, dehydration, and cyclization; the exact pathway is ultimately dependent on the reaction conditions but

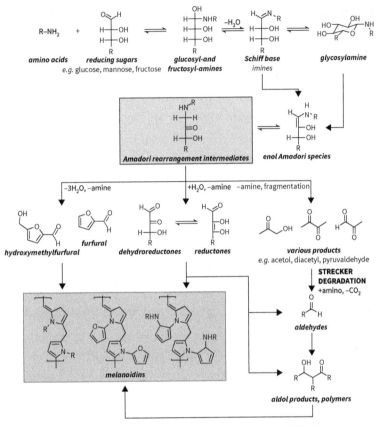

Figure 4.4 An overview of Maillard products in roasted food. R = various hydroxylated alkyl substituents.

Image adapted from Vernon and Parkanyi (1982) and Perez-Locas and Yaylayan (2010).

undoubtedly is responsible for the wide range of flavor molecules expressed in coffee (Jousse et al., 2002).

The degree of roasting depicted in Figure 4.4 may also be measured by key chemical indices. For example, using reversed-phase high-performance liquid chromatography with a diode array detector, R-dicarbonyl content can be examined in green coffee and during various roasting stages (Daglia et al., 2007). Minimal amounts of glyoxal and methylglyoxal were found to occur naturally in green coffee, increasing rapidly to peak in the early stages of roasting, then steadily declining. Diacetyl, however, was only found in highly developed coffee beans in low concentrations. As a result, it was concluded that light- and medium-roasted coffees (i.e., short and medium development, respectively) with high concentrations of glyoxal and methylglyoxal could be distinguished from dark-roasted coffees with lower concentrations of glyoxal and methylglyoxal, as well as a small diacetyl content.

One notable study examined chemical profiles of volatile fractions obtained using headspace solid-phase microextraction (SPME) gas chromatography–mass spectrometry, identifying several compounds that were discriminative of bean color and coffee variety (Ruosi et al., 2012). Of the techniques used to examine headspace, SPME has been shown to be efficient for volatile flavor compounds, contingent on the selection of an adequate fiber (Roberts et al., 2000). Notably, the compounds determined to be exclusively indicative of roasting include 2-methylfuran, 2,3-butanedione, 2,3-pentanedione, pyridine, 3-ethylpyridine, furfural, 5-methylfurfural, and guaiacol. The ratio between 5-methylfurfural and 2-acetylfuran was determined to yield the best linear correlation with the color obtained at different roast conditions. Compounds indicating both roast and coffee variety include 2,3-butanedione, 2,3-pentanedione, 1-methylpyrrole, pyridine, furfural, 5-methylfurfural, 1-furfurylpyrrole, guaiacol, and acetylpyrrole. Compounds determined to be indicative of coffee variety include 2-vinyl-5-methylfuran, 2-propylpyrazine, and 2-acetylfuran. The concentration of furan that forms as a result of the thermal degradation of sugars and amino acids and the thermal oxidation of ascorbic acid and polyunsaturated fatty acids (Perez-Locas and Yaylayan, 2004) is linearly correlated with roast color—darker beans contain higher furan levels (Arisseto et al., 2011).

(Poly)Phenols are widely recognized bioactive antioxidant and radical scavenging compounds (Samsonowicz et al., 2019). Of the common phenolics, chlorogenic acid and isoflavones have recently been identified in varying concentrations (Alves et al., 2010). Over the course of the roasting process, the overall concentration of isoflavones decreased, which is further reflected by the elevated concentration of isoflavones in lighter coffees. Additionally, some phenols and N-heterocycles are thought to impart a bitter taste in coffee—a flavor often considered undesirable. The formation of bitter-tasting compounds as they relate to roasting degree and time was studied (Blumberg et al., 2010). Both the roast temperature and time significantly impact the concentration of the identified bitter compounds in standard coffee brews; increased roast gas temperature resulted in an increased intensity of bitter tastants, while increased time resulted in decreased concentrations of monocaffeoyl quinic acids, monocaffeoylquinides, and dicaffeoyl-quinides, with a subsequent increased concentration of 4-vinylcatechol oligomers.

The evolution of chemical compounds present within the bean throughout the roasting process is indicative of roasting degree and can inform final cup quality based on the presence or absence of specific compounds. For example, chlorogenic acids represent the majority of polyphenols in green coffee beans and are associated with an astringent taste. Throughout the roasting process,

ARTISANAL COFFEE

these acids are thermally degraded, forming quinic acid, cinnamic acid, and quinide. Decreasing the chlorogenic acid content is one of the reasons darker coffees (i.e., those that have been roasted for a longer period of time) are associated with reduced astringency (Wei et al., 2012). This must be balanced, however, with the formation of quinic acids and quinides, which are also contributing factors to the bitterness of coffee. Chlorogenic acid content, as measured by the concentration of caffeoylquinic acids, dicaffeoylquinic acids, and feruloylquinic acids, is correlated with the perceived bitterness and acidity of coffee (Figure 4.5) (Moon et al., 2009).

Flavor and Aroma Descriptions

The roasting process generates a multitude of compounds in concentrations that depend on the aforementioned processing. There are obvious challenges associated with attributing the quality and quantity of these compounds to linguistic descriptors. Instead, World Coffee Research has facilitated our description of coffee flavors by establishing the Sensory Lexicon (World Coffee Research, 2017), which provides a quantifiable, descriptive, and replicable metric by which a cup of coffee may be assessed. The Sensory Lexicon is associated with the Coffee Taster's Flavor Wheel (Figure 4.6), which details some familiar flavor notes perceived in coffee (Spencer et al., 2016). Each flavor descriptor has an associated reference recipe, enabling interested readers to train themselves to taste. For example, the reference for both aroma and taste of "blueberry" can be prepared by smelling/ tasting a sample of Oregon Fruit Products Blueberries in Light Syrup (canned).

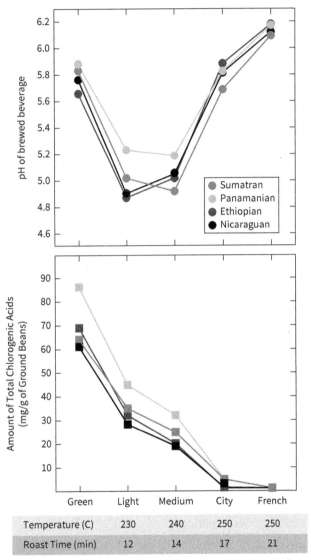

Figure 4.5 Roasting conditions dictate the final chlorogenic acid content and pH. While there exist slight discrepancies, pH and chlorogenic acid content are essentially independent of bean origin.

Figure generated using data presented Moon et al. (2009).

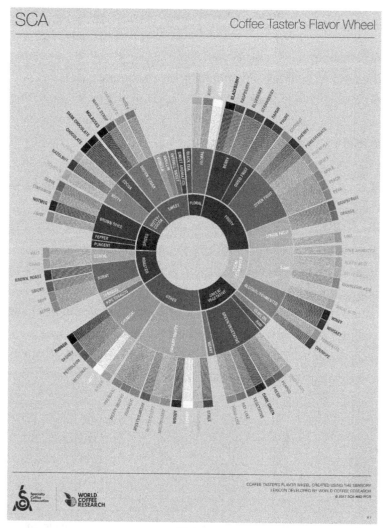

Figure 4.6 Specialty Coffee Association Coffee Taster's Flavor Wheel.
A tool used to homogenize language associated with tasting and
describing coffee, using words and standard flavors presented by World
Coffee Research.
Image reproduced from Spencer et al. (2016).

Beyond flavor notes, however, the monetary value of a coffee must be objectively assessed so that coffee producers can negotiate reasonable prices for their crops. Coffee is purchased by first tasting the product and then either bidding on an allotment or direct sale. Coffee is generally graded on a scale out of 100 points using a scoring sheet. One example is presented in Figure 4.7. The internationally agreed-upon method for assessing and grading coffee quality is called "cupping." In this method, a standardized amount of coffee is ground and submerged in water. After a nominal steeping time, the coffee is assessed over several minutes, with higher scores in the categories indicating higher value coffees. Specialty coffees are defined as those scoring above 80 points across five identical cups, assessed using the cupping protocol. A coffee scoring at least 80 points will primarily feature a distinct balance between each evaluation rubric while also being free of any defective or undesirable qualities. Monetary value in most cases rises linearly with quality score, where super-premium coffees scoring upward of 90 points are as expensive as they are exciting and rare.

Roast Level	Fragrance/Aroma		Flavor	Acidity	Body	Sweetness	Overall
	╷╷╷╷╷╷╷╷╷ 6 7 8 9 10		╷╷╷╷╷╷╷╷╷ 6 7 8 9 10	╷╷╷╷╷╷╷╷╷ 6 7 8 9 10	╷╷╷╷╷╷╷╷╷ 6 7 8 9 10	☐☐☐☐☐ Clean cup	╷╷╷╷╷╷╷╷╷ 6 7 8 9 10
	Dry	Crust	Fragrance/Aroma Quality	Intensity ⬜ High	Intensity ⬜ High	☐☐☐☐☐ Uniformity	Defects (subtract)
			Aftertaste ╷╷╷╷╷╷╷╷╷ 6 7 8 9 10	⬜ Low	⬜ Low	☐☐☐☐☐ Balance ╷╷╷╷╷╷╷╷╷ 6 7 8 9 10	☐ ✕☐ ☐ # of cups Intensity Taint = 2 Fault = 4
Notes							Final Score

Figure 4.7 Specialty Coffee Association coffee quality cupping form. Coffees are graded on a 100-point scale, and specialty coffees are defined as those that score above 80 points. Note that the scale ranges from 6 to 10, where 6 is "good" and 10 is "even better than outstanding," inferring that all coffee is good.

Specialty Coffee Association Arabica Cupping Form (2004–2023), designed and intended to be used in conjunction with the *SCA Protocol for Cupping Specialty Coffee (2004–2023)*, is reprinted with permission from the Specialty Coffee Association (https://sca.coffee/cupping).

Cupping Protocol

Cupping is the internationally agreed-upon method for assessing coffee quality. First, ~0.05 g of roasted coffee per milliliter of water should be ground coarse and placed in a vessel of ~250 mL. Hot water (200 ± 2°F) is then added to the coffee and left unperturbed for 4 minutes. A crust of floating coffee grounds is then disturbed with a spoon in a few gentle pushes through the surface. Remaining floating coffee grounds are discarded using two spoons to scrape/skim the surface. The coffee is then tasted using a spoon.

Coffee Preparation

Moving from whole, roasted beans to a hot cup of coffee involves two final steps: i) grinding and ii) extraction/brewing. Altering both grind size and extraction procedure provides added control over the quality of the final cup. Grinding specifically serves to increase the accessible surface area, while also playing a significant role in determining flow. In the former, the diffusion distance for soluble compounds during extraction is reduced with reducing grind size (Spiro and Selwood, 1984; Cordoba et al., 2019). The rate of extraction and the relative populations of the various compounds in coffee then depend on tangible variables such as water temperature and roast profile and are dictated by the polarity of the compounds in coffee (López et al., 2016; Bhumiratana et al., 2011). As one could imagine, there is a delicate balance between the grind size and distribution and the brewing parameters.

Grinding roasted coffee beans releases volatile organic compounds (VOCs) that were formed during the roasting process, but determining which VOCs are responsible for nuanced aroma remains challenging. This is mainly due to the sheer number of unique compounds that have been identified to date (~900 [Mateus et al., 2007a, 2007b]). Particle size has been shown to dictate the release kinetics of VOCs (Wang et al., 2016); smaller particles lead to faster release kinetics, which may be due to the higher surface area and porosity. A separate study demonstrated that coffee grounds have variable porosity, depending on particle size (Mateus et al., 2007a, 2007b). Together, these studies highlight that there is more work to be done to understand the role of porosity and density, particularly given the increased focus on differences in coffee origin and roast profiles.

Neither manual nor automatic coffee grinders, however, produce unimodal particle size distributions. When measured on a laser diffraction particle size analyzer (Figure 4.8), ground coffee exists in two families of sizes: fines (particles <100 μm) and boulders (particles >100 μm). Initial studies indicate that, for a fixed grinder setting, particle size distribution is independent of roast

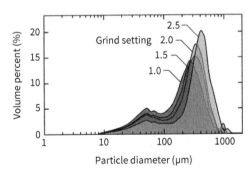

Figure 4.8 The particle size distribution of a representative specialty coffee ground on a Mahlkönig EK 43 at the grind settings shown. Volume percent is measured using a laser diffraction method presented in Uman et al. (2016).

development (Uman et al., 2016). The radius of fine particulates was found to depend on bean temperature, and coffee professionals have begun using "bean temperature-while-grinding" as a variable to achieve more uniform particle size distributions.

Grind setting must be matched to brew method. Coarse grind settings (mode boulder = ~1000 μm) are generally used for systems involving long (~4 minutes) water–coffee contact time. For faster brewing methods, in particular those that lower bed permeability to achieve elevated pressure, grind settings tend to have mode boulder sizes closer to 300 μm. However, flow restriction and overall sensorial quality also require careful control of the *brew ratio*, the ratio of dry coffee mass to water mass used during the brew. When combined, the brew ratio and grind setting are the two central coffee variables that must be controlled to achieve an artisanal beverage.

The brew ratio is one of the key determinants for coffee strength (concentration) and quality (Andueza et al., 2007). In the coffee industry, strength is used to discuss the concentration of total dissolved solids (TDS) in liquid coffee, which is reported as a percentage mass/mass. At the time of writing, strength is measured using the refractive index, converted to coffee TDS empirically (see Fedele, 2012; Batali et al., 2020). TDS is often intrinsic to a particular brew method, with methods like pour-over producing coffee approximately 1.3%–1.5% TDS (weaker) and espresso with 8%–12% TDS (stronger). Even within one brew method, both grind setting and brew ratio may be used to control strength. By way of example, some consumers may prefer a less concentrated, more delicate cup of coffee at 1.3% TDS, while others would prefer a heavier (both physically and sensorially), more hearty cup at 1.8% TDS. Neither option is viewed as bad within the industry, assuming that both are prepared so as to not extract negative flavors from the grounds. The most practical method to achieve a specific TDS is by making adjustments to the water-to-coffee ratio. Increasing or decreasing the input of ground coffee while keeping the water input fixed, respectively, increases or decreases the TDS. Although changing the brew ratio can in principle also affect the extraction of compounds near equilibrium, there is no evidence to suggest that any compounds in coffee exist near their solubility limit in water (Bradley and Hendon, 2017).

As a counterpoint to the subjectivity of strength, there are good reasons to choose a desired strength. Roast degree is the primary factor to take into account when determining at what strength a particular coffee ought to be prepared to best showcase its flavor potential. As discussed in the roasting section above, the evolution of chemical compounds that produce both negative and positive flavors is indicative of roasting degree. Depending on how developed the coffee is, specific water-to-coffee ratios can help hide the prevalence of negative flavors and heighten the positive ones.

Generally speaking, a lighter-roasted coffee will have maintained the potential for more delicate flavors, like perhaps rose, pineapple, and honey (per Figure 4.6). But we also know that lighter roasts retain the potential for astringent qualities, attributed to a greater presence of polyphenols (Król et al., 2020). By choosing to increase the water-to-coffee ratio by reducing the ground coffee input, the TDS is lowered and the astringent compounds are subdued, allowing for improved perception of the positive and more delicate flavors. On the other hand, a darker-roasted coffee will have developed the potential for heartier flavors. These chocolate, clove, and raisin notes are competing with higher concentrations of quinic acids and quinides that contribute to heightened bitterness. Thus, by choosing to narrow the water-to-coffee ratio by increasing the ground coffee input, the TDS is increased and the bitter compounds become muted beneath the improved perception of the more desirable flavors.

Water Chemistry

While some have tried using non-aqueous solvents to extract from coffee (Efthymiopoulos et al., 2018), water is generally used as the solvent. Initially, roasted coffee contains very little water and proceeds to swell during brewing. The water affects the coffee in several ways: by acting as a solvent for organic material (i.e., the flavor), by suspending particles (adding to both the extraction and mouthfeel), and by facilitating the release of volatile compounds from within the bean. Scanning electron microscopic images revealed that water enters the coffee and produces vacuoles of trapped gas (Mateus et al., 2007a, 2007b). Mechanistically, it is thought that gaseous and volatile compounds are displaced by water, resulting in an effervescent appearance referred to as "blooming" and the dissolution of a portion of the gas molecules and volatiles in the water. The speed of this stage is primarily impacted by the grind size/shape, particle density/distribution, and bean geographical origin/species (Spiro, 1993). For example, Robusta coffee beans often exhibit higher water absorption compared to Arabica beans, resulting in a longer wetting stage and subsequent overall extraction time (Ludwig et al., 2012).

Once wet, extraction of soluble products begins. While there may be other competing factors at play, the process is generally governed by diffusion. According to Fick's law, diffusion is a function of the time, temperature, liquid viscosity, layer depth, and solute concentration in the liquid and the solid. Toward the latter, it is well understood that local concentration gradients play a significant role in mass transport. In an analogous heterogeneous system—the interface between electrode and solution in an electrochemical measurement—stirring/agitation overcomes the mass transport (or diffusional) limits (Houghton and Kuhn, 1974), and the same applies in coffee extraction. Brew methods that involve unstirred contact between all water and coffee (i.e., immersion brewing) may be affected by mass transport. Most other brew methods are limited by other variables such as contact time and temperature.

However, soluble minerals in water also play an operative role in structuring the flavor of coffee extractions. Dissolved ions, such as transition metals, halides, nitrate, sulfate, and carbonate, are thought to each serve a different role (Lockhart et al., 1955; Pangborn et al., 1971). A theoretical study examined the binding energy of three cations (Na^+, Mg^{2+}, Ca^{2+}) commonly found in water and seven representative coffee compounds (lactic acid, malic acid, citric acid, quinic acid, chlorogenic acid, alkaloid caffeine, and eugenol) (Hendon et al., 2014). There, a general conclusion was made that as ionic strength increases (Solomon, 2001), so should the propensity for water to extract more polar compounds. Navarini and Rivetti (2010) examined the role of other ions in coffee brew water, highlighting that increased alkalinity (the cumulative measurement of base within the solution) had an impact on the sensory properties of espresso-based coffee. Additionally, increased bicarbonate resulted in an increased appearance and persistence of the characteristic espresso foam layer, the crema (directly correlated with the amount of available CO_2). Yet, while the initial foam is denser (as measured by a higher foam index), it dissipates faster than water containing less bicarbonate. Colonna-Dashwood and Hendon (2015) offered a water brew control chart (Figure 4.9), which summarized both literature and anecdotal findings of how coffee flavor can be predicted based on mineral composition of brew water. It should be noted that mineral-rich water can have adverse effects on brewing equipment. Harder (i.e., high mineral content) water may result in the formation of $CaCO_3$. The role of buffering agents in water (i.e., water alkalinity), such as bicarbonate, is also important considering the desirable acidic components of many light roasts.

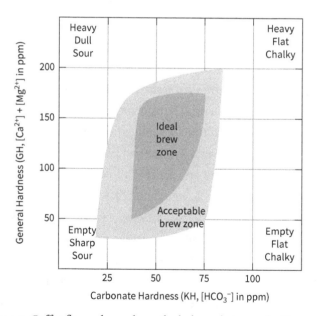

Figure 4.9 Coffee flavor depends on the balance between buffer concentration (HCO_3^-) and divalent metal ions. An excess of either will have negative impacts on coffee flavor.
Image used with permission from Colonna-Dashwood and Hendon (2015).

Brewing Methods

Brewing procedure mechanics range from simple stovetop boiling vessels to complex glassware made to siphon hot water to ground coffee. While seemingly diverse, all extraction techniques fall into one of two classes and are considered either immersion or infusion/flow brewing techniques. Immersion methods allow grounds to contact the full amount of water at any given time throughout the brewing process (e.g., French press, AeroPress). In contrast, infusion/flow methods are associated with a flow of water, driven by either gravity or a pressurized system (e.g., pour-over, espresso).

Immersion

Immersion methods are among the oldest coffee brewing methods and, thus, the simplest. It is primarily for this reason that cupping was chosen as the international standard for quality evaluation; its simplicity and replicability are unmatched. The main differences between immersion brewing methods are the grind size and brewing apparatus. All of these methods, however, involve a mixture of water and coffee grounds. In fact, even more so than cupping, the simplest extraction method is boiled coffee, in which coarse coffee grounds are mixed with water and brought to boil. Often, a filter is not used to prevent the grounds from flowing into the cup; rather, the grounds are allowed to settle during cooling. The size of the grind, indeed, influences the extraction rate and yield. Among the most common immersion methods are Turkish coffee, French press, and Toddy-style cold brew.

French press is perhaps the most prevalent immersion method in the United States. This could be due to its ease of usability but also in part to the uniquely textured cup of coffee that it produces. Here, coarsely ground coffee and hot water are combined in a tall, cylindrical vessel fitted with a metal filter attached to a plunger. At the end of a relatively long brewing time (~4 minutes), the plunger is driven down, in principle preventing the grounds from being poured into the final cup. The texture, referred to as "mouthfeel," of a French press coffee is heavier on the palate, owing to the higher turbidity compared to other brew methods. This enhanced mouthfeel is the result of using a metal filter, which allows some fine insoluble particles to pass into the resulting beverage. However, too much particulate can result in an unpleasantly gritty and bitter coffee, which is why coarsely ground coffee is necessary to mitigate the amount of insoluble material that is allowed to pass into the cup. However, brewing with coarsely ground coffee will always pose challenges as it works against extraction efficiency. Boulders have a lower surface area-to-volume ratio and therefore an increased diffusion distance to extract soluble compounds from. Using a longer brewing time helps maximize coffee solid dissolution (Spiro et al., 1989) and achieve optimal extraction yield (EY). Additionally, coarser grounds are especially absorbent; and without the benefit of atmospheric pressure, a greater amount of liquid remains trapped in the grounds even after decanting, which unavoidably subtracts from potential strength and EY. For this reason, it can be beneficial to use slightly more ground coffee when making French press coffee to achieve the same TDS and EY as an infusion/flow method using the same amount of water. While a 1:16-g coffee-to-water ratio is reasonable for filter coffees, a 1:13 ratio may be more optimal for French press. It is also worth noting that the added pressure on the plunger may work to extract extra solids, which is why some references have classified this as a pressure method, although this is certainly not its intended purpose. The AeroPress, however, is an example of an immersion brewer that uses a finer grind setting, a shorter brew time (~1.5 minutes), and manually applied pressure.

Infusion/Flow

Infusion processes flow water *through* a coffee bed, resulting in a local chemical environment where new water is meeting the coffee surfaces. Within this process, there are two further subcategories: those that use atmospheric pressure and those that do not. Examples of atmospheric infusion processes include pour-over coffee apparatuses: V60, Chemex, and so forth. Non-atmospheric examples include espresso and the Moka (Bialetti) pot. The inclusion of pressure and a heavily reduced brew ratio (1:2 coffee to water) results in both espresso and Moka pots achieving very higher concentrations (in the vicinity of 8%–12% TDS), typically unachievable in pour-over and immersion approaches. In a professional setting, consistency of preparation is controlled by a barista, as is quality and intentionality. Additionally, there are many esoteric tools that can be used to help achieve better brew quality. It is easy to get distracted by technologies in coffee, but fundamentally all coffee professionals focus on four primary controls: uniform and controlled grind size, coffee-to-water ratio measured by weight, controlled water temperature, and intentional water application.

Unlike other brewing methods, the brew parameters for espresso are historically defined. The Specialty Coffee Association states that an espresso is a 25–35-mL (~20–30 g) beverage prepared from 7 to 9 g of ground coffee made with water heated to 92–95°C, forced through the granular bed under 9–10 bars of static water pressure and a total flow time of 20–30 seconds. However, modern coffee shots tend to favor using more coffee, ~15–20 g of ground coffee, to prepare a beverage of ~35–60 mL in volume. This cultural shift is in part due to the improved agricultural practices that have led to highly desirable flavors, and modern brew protocols have been adapted to target these flavors by roasting lighter but increasing the brew ratio to maintain beverage strength. Work by Ellero and Navarini (2019) and Cameron et al. (2020) revisited espresso preparation from a modeling perspective, and both concluded that fine particulates are thought to govern flavor in espresso. The definition of modern espresso is moving away from the rigid parameters detailed above, instead being characterized by its concentration (>8% TDS) and the equipment in which the drink was produced. This divergence from history is due to the emergence of highly programmable espresso machines, with sophisticated pressure profiling and water temperature, paired with advances in grinders, agricultural advances in coffee quality, and other minor variables (tamp force, espresso basket geometry, etc.).

While French press may be the most prevalent immersion method in the United States, filter coffee might be the most common overall. Filter coffee utilizes a simple infusion process that relies on atmospheric pressure. The two most common preparation styles of filter coffee are the automatic electric coffee maker and the manual pour-over. The pour-over method is simple in concept—hot water poured through ground coffee contained in a basket-shaped apparatus with a filtration medium—but the variations in techniques, apparatuses, recipes, and results are seemingly limitless. In order to achieve an intended outcome, a professional barista must control for almost every variable affecting extraction. Atmospheric pressure during pour-over brewing is a constant variable that is seemingly out of a barista's control as it relentlessly pulls water beyond its crucial contact with coffee grounds. But this is not actually the case; there are techniques a barista can employ to make percolation work for them. Manipulating the grind setting is often the first approach, where a coarser grind allows percolation to occur faster and a finer grind to slow the rate of flow. But there is only so coarse or fine a grind that can be taken before extraction quality suffers and flavor sacrificed. Similar to espresso preparation, a uniform grind size is also very important, not only to prevent uneven extraction but also to prevent the migration of very fine particles to the bottom of the basket that can effectively clog the pores of the filtration medium. A tool ubiquitous to any cafe preparing pour-over specialty coffee is a water kettle with a long and narrow S-shaped spout called a "gooseneck kettle," among other names. The narrow spout focuses the hot water into a laminar stream, allowing the user to guide the flow intentionally and delicately over the ground coffee. Flow rate is also easily controllable using a gooseneck kettle. The coffee's contact time with water can be increased by pouring slowly, sometimes extremely so. Frequent pauses in pours can also help extend the contact time. Sequential pouring may also help maintain the ideal brewing temperature of the water–coffee mixture as each new pour replenishes some lost heat. Though seemingly random, these pouring phases can be a precisely choreographed technique. A professional barista can time their pours so precisely that the very last drop of cup after cup of coffee falls within 10 seconds of a total expected brew time. Each practiced manipulation of any given brewing variable is made with the sole expectation to extract the most balanced coffee possible.

The difference between brewing procedures mainly lies in the rate of extraction of molecules from coffee grounds into water. These kinetic processes are tuned using brew time, temperature, pressure, grind size/distribution, and vessel geometry to extract optimal levels of tastants, odorants, and antioxidants. For example, automatic coffee machines are characterized by an increased ratio of coffee grounds to water, which results in a decreased dilution factor—leading to a more intense cup (Arisseto et al., 2011). Similarly, the presence of antioxidants in brewed coffee may be tuned by the extraction method (Ludwig et al., 2012). Interestingly, extracted antioxidant concentration is correlated with extraction time; higher levels of antioxidants were extracted throughout a filter brew, while antioxidant extraction decreased with time in espresso brews. This can be understood from thermodynamics—a kinetically limited exothermic solvation should show increased population with increased time but decreased population with increased temperature.

Generally, contact time is a dependent variable in the sense that it varies based on brew ratio, grind setting, and water pressure. However, time can be controlled in certain brew methods, and extended contact time is thought to increase perceived bitterness. Also, increased contact between hot water and coffee may induce hydrolytic release of caffeoylquinic acids, the most abundant chlorogenic acid in coffee (Kang et al., 2019). At minimum, higher brew temperatures result in higher phenolic content, indicating that the solubility of larger organics is likely endothermic.

Beyond chlorogenic acids, furans also contribute to the overall flavor profile and have been identified as the major volatile component in both cold and hot brew (by pour-over method) coffee (Kang et al., 2019; Crews and Castle, 2007). Further demonstrating the importance of brew temperature, 2-methylfuran (associated with fruity flavor notes) was only identified in hot brews. This is likely a kinetic effect, considering that 2-methylfuran has the lowest boiling point of the 19 furans that were identified in this study. Thus, over the course of the extended brewing time for cold brew coffee (12 hours), either the 2-methylfuran evaporated or the equilibrium constant of solubility is heavily temperature-dependent (the latter is true but less likely). Beyond temperature, use of a filter, as implemented in French press brewing, decreases final furan levels, relative to the conventionally filter-less Turkish coffee method (Amanpour and Selli, 2015). In the latter, it is likely not the filter that dictated the furan levels but rather the tremendously different brew parameters used in French press versus Turkish preparations.

Brew basket geometry may also play a role in the final sensory quality of the coffee due to the altered coffee particle–water interaction time and the pressure/flow of the water as influenced by the shape of the brew basket (Frost et al., 2019). Coffee brewed using semi-conical and flat-bottom baskets exhibited differences in smoke aroma, sweetness, and tobacco flavor, as assessed by Q graders; discrepancies were believed to originate from the higher percentage of TDS in the semi-conical brewed cup over the flat-bottom brew. This study, however, neglected to control for TDS, one of the primary factors in cup experience. Thus, it is unsurprising that the examined basket geometries resulted in differing sensorial experiences.

Beyond hot brewed beverages, cold brew coffee has seen a recent increase in popularity, owing to the alternative flavor profile present as a result of the room temperature and extended brewing procedure. Within this process, coffee granules are steeped for an extended period of time (~8–24 hours) at room temperature or below (20–25°C). Furfural was found to be the major volatile component in cold brewed coffee and contributes to the sweet, bread-like, and caramel flavors (Casal et al., 2015). Pyrazines, offering nutty, earthy, roasty, and green notes, are more prevalent at higher serving temperatures yet are the second most prevalent volatile compound (Cordoba et al., 2019). Many of the flavor and odor compounds identified in the cold brew coffee beverages are reminiscent of those found in hot beverages, the major difference being the concentration, due to the temperature and extended extraction time of the brewing process.

Thinking more about espresso, however, temperature has been shown to linearly correlate with the concentration, EY, and TDS. For example, the lowest output temperature was exhibited by the bar machine and corresponded to the lowest concentration, EY, and TDS (Parenti et al., 2014). Thus, a better-insulated brew chamber would decrease temperature loss over the brew process—which may impact the final cup volatile concentration. Espresso, a polyphasic system, is characterized by both its intense flavor profile and distinguishing crema, which is responsible for the smoother mouthfeel over other brew methods and contributes to the timely release of volatiles from the brew. The crema is composed of fine gas bubbles sitting on an emulsion of coffee oil droplets suspended in an aqueous solution (Ferrari et al., 2007). Interestingly, density and viscosity of the espresso brew procedures were consistent, even though capsule systems exhibited higher total lipid content compared to conventional espresso. Typically, viscosity and total lipid concentration are logarithmically correlated (Kilcast and Clegg, 2002), playing an important role in the mouthfeel and creaminess of the final brew.

Bean freshness (Wang et al., 2019) and lipid content also dictate the foam indices and persistency (with both galactomannan and arabinogalactan being the most abundant polysaccharides found in coffee brews [Moreira et al., 2015]) and are primarily responsible for the viscoelastic nature of the interface between the crema and the espresso brew. While capsule systems are associated with higher foam indices and persistence, bar machine brews are characterized by more intense flavor, as suggested by the higher concentration of aromatic compounds. The authors suggest that the foam density is negatively correlated with the concentration of VOCs in the headspace (Kilcast and Clegg, 2002); the presence of the foam structure may work to prevent the evaporation of VOCs, thereby resulting in lower measured concentrations above capsule brews associated with a higher foam index (Dold et al., 2011).

The optimal brewing procedure and coffee selection to perfect the crema is a well-studied area, with the perfect crema persisting for 2 minutes after brewing and contributing to 10% of the espresso volume (Illy and Viani, 2005), which may depend on the origin and roast of the bean (Maeztu et al., 2001). A wide array of chemical phenomena and compounds are responsible for the formation and persistence of the crema. Indeed, entire books have been written on espresso, and an excellent review of crema formation and stability may be found (Illy and Navarini, 2011). More broadly, crema formation depends on the protein composition of the bean, as well as the protein derivatives formed as a result of the Maillard reaction (e.g., melanoidins) (see Nunes et al., 1997).

Preservation and Storage

Preservation and storage of coffee beans, grounds, and brewed beverages are pivotal to ensuring cup quality; loss of quality is commonly associated with augmented acidity, sourness, bitterness, and/or astringency (Pérez-Martínez et al., 2008). In short, staling is predictable, but the resultant flavor is not. Storage considerations are essential from the moment the raw coffee bean is picked to the final brew. In fact, green coffee beans can be stored as long as 3 years prior to roasting (Bucheli et al., 1998), during which fermentation reactions alter the aromatic profile as a function of the storage temperature and water content (Scheidig et al. 2007). Specialty roasters tend to not have this problem as they turn their stock over seasonally. However, independent of the coffee quality, aromatic changes are minimized at lower storage temperatures and decreased water content. While the storage conditions of green coffee beans are important, often this is not within the control of the final coffee consumer. Thus, we will focus on the impact of storage conditions on the chemical composition of roasted coffee beans and beverages but emphasize that cryogenics of green coffee is a developing area that a handful of roasters are already experimenting with.

Green coffee contains approximately 12% water by mass. Water activity is an important parameter in food safety and preservation and essentially measures the chemical potential of water within a food matrix. Recently, water activity and moisture content were monitored by a green coffee importer, Café Imports (Fretheim 2019). Water activity levels higher than 11.5% may lead to off-flavor notes, which Café Imports attributes to lipid oxidation or a Maillard spoilage reaction. The effect of these reactions was observed through a tenuous correlation between water activity and shelf degradation, where coffees with lower initial water activities lost fewer cupping points and were less volatile.

Due to the high roasting temperatures, roasted coffee is essentially anhydrous and therefore has very low water activity. This, coupled with the presence of antimicrobial Maillard reaction products, protects roasted coffee beans against enzymatic and microbial spoilage (Anese et al., 2006). These characteristics, however, do not prevent degradation of volatile compounds and oxidation reactions, which are known to negatively impact the sensory profile of coffee brews. Imminent degradation of critical flavor and aroma compounds, referred to as "coffee staling," and development of "off-flavor" notes by oxidation reactions may be avoided by optimizing preservation and storage conditions. In light of the hydrolytic and oxidative mechanisms of coffee staling, it is noted that both caffeine and trigonelline compositions are independent of coffee storage time and temperature in the presence or absence of oxygen (Pérez-Martínez et al., 2008).

One of the negative flavor notes attributed to coffee staling is an increased sourness, which describes sharp, biting, vinegar flavor notes and is not related to the commonly enjoyed acidic qualities of lighter roasts. is correlated with an observed decrease in brewed pH (Manzocco and Nicoli, 2007). Increased acidity is due to a variety of reactions, including hydrolysis of lactones and esters as well as non-enzymatic browning reactions that occur even after the completion of the roasting process (rate is decreased by the impeded mobility of reactants within the bean as a result of the cooling process). Indeed, the rate of H_3O^+ formation is directly related to temperature, following an Arrhenius-type relationship for coffee samples exhibiting water activities higher than 0.52. Specifically, water activity impacted both the calculated activation energy and the frequency factor of the Arrhenius equation, indicating an overall impact on the temperature sensitivity of stored coffee.

The importance of storage temperature on the rate of H_3O^+ formation decreases with increasing water activity due to the increased molecular mobility of the compounds within the coffee. Increased water activity improves the mobility of reactants within the grounds, which is otherwise kinetically impeded at lower water activity levels. Alternatively, increased acidity due to staling product formation might be mitigated by brewing with water containing higher levels of buffering agents, such as bicarbonate ions; while no studies have pursued this to date, we see this as a viable research objective to improve the sensory quality of stale coffee through control of water chemistry.

The presence of oxygen is an important consideration for home brewing coffee, where packaged beans and grounds are repeatedly introduced to the atmospheric environment. Owing to the presence of polyphenols and antioxidants, the coffee lipid fraction is associated with a high oxidative stability (Nicoli et al., 1997). Therefore, repeated exposure of stored coffee to an oxygen environment (e.g., opening and closing of coffee containers) may, indeed, impact the overall flavor profile due to oxidative reactions with the Maillard reaction products.

Retention of volatile compounds in coffee is also important to ensuring aromatic quality. While off-gassing is a spontaneous process (Smrke et al., 2018), volatile compounds do significantly contribute to the sensory properties of brewed coffee. Increasing internal bean water content results in loss of important volatile compounds because water is thought to compete for the polar and non-polar sites that the volatile compounds are bound to within the coffee matrix (Anese et al., 2006). Loss of volatiles is primarily a thermodynamically driven process and was found to follow a pseudo-zero-order kinetics model, enabling the development of a secondary shelf-life prediction model.

The unstable aromatic compounds contributing to the roasty–sulfury smell, namely 2-furfurylthiol, of fresh coffee brews necessitates beverage storage considerations because of its rapid concentration decrease upon processing and storage (Müller and Hofmann, 2007). This is of specific concern for canned coffee beverages and instant coffee grounds. Oxidative coupling of 2-furfurylthiol to hydroxyhydroquinone, yielding several phenol/thiol conjugates, is responsible for the immediate degradation of the fresh coffee roasted aroma. Studies focusing on preventing the storage-induced degradation of the critical roasted aroma of fresh coffee are ongoing.

Freezing of both whole bean and brewed coffee offers desirable environmental parameters for improved aromatic and flavor profiles of stored brews. Three freezing methods (vacuum, air, and contact) were investigated to identify optimal storage conditions for coffee extracts (Silva and Schmidt, 2019). Convection and conduction control heat transfer in conventional (air and contact) freezing methods; thus, relevant parameters influencing the rate of freezing include sample thickness, in addition to the temperature and rate of air flow. Contrastingly, vacuum freezing, due to the low pressure, induces water evaporation and subsequent sublimation of formed ice crystals—thereby resulting in freezing and drying of the final product to be achieved in one process step. Scanning electron microscopy allowed the microstructure of varying concentrations of coffee extracts to be evaluated as a function of freezing procedure. Unsurprisingly, higher moisture content yielded microstructures with increased porosity; maximum porosity was achieved with vacuum freezing, which may be attributed to boiling induced by the evaporative cooling step.

Like thermally treated milk, ultra-high temperature (UHT) coffee has been shown to extend the shelf life of coffee brew and increase the beverage stability (Sopelana et al., 2013). In contrast to typical pasteurization and sterilization techniques, UHT is a shorter thermal treatment, which may not modify the aromas and flavors of coffee beverages during storage. Interestingly, UHT-treated coffee brews at 110°C and 120°C were closest sensorially to untreated coffee brews, although UHT-treated brews were still associated with a slight astringency and aftertaste flavors; and this is an ongoing issue for prepackaged cold brew products.

Outlook and Actionable Information

Establishing a relationship behind final cup quality and the coffee production process necessitates a fundamental understanding of the physical and chemical phenomena behind the coffee journey from plant to cup. Without such scientific, data-centric endeavors, only a qualitative understanding is afforded—preventing important advances in the coffee industry, which may work to improve the sustainability of coffee production and consumption.

Here, we aimed to broadly canvas coffee research to date, while highlighting the key artisanal aspects of its preparation. In doing so, we conclude a few certainties.

1. The efficacy of novel advances must all be measured by the impact (positive or negative) on the final cup quality. Indeed, improvements in agricultural methods and technologies surrounding the proliferation of coffee plants may not be fully appreciated by the final consumer but would contribute to the economic health of coffee-producing nations.

2. Theoretical and experimental studies on the fundamental chemistry and physics behind flavor and aroma compound generation and extraction offer avenues that will enable new paradigms in coffee roasting—the critical stage for flavor generation.

3. Brew methods play an operative role in accessing flavors generated during the roast, but advances in post-roast coffee treatments (including water chemistry) offer avenues to further modulate the coffee flavor. One can imagine the development of a material that selectively absorbs tainted flavors or even biochemical approaches, such as serine-type carboxypeptidases produced by *Aspergillus oryzae*, a filamentous fungus, to reduce bitterness (Murthy et al., 2019).

4. The expertise offered by a professional barista is deeper than one may assume. Our original conjecture, that coffee is dissimilar to other artisanal beverages, should now be

validated. Baristas harness excellent laboratory technique to carefully prepare coffee beverages from ever-changing coffee beans.

5. There are still many unanswered questions, and the seasonality and variability of coffee complicate our ability to develop a fundamental understanding of the physical and chemical considerations that give rise to the qualities in the cup.

In summary, the scope of this perspective has been largely focused on bean selection and beverage preparation for improved coffee experience. There are countless other interesting topics we did not explore in great detail, including benefits beyond human consumption (McNutt and He, 2019; Toschi et al., 2014); for example, spent coffee grounds have been explored as beneficial components in dyes (Bae and Hong, 2019; Schroeder et al., 2019), energy storage devices and fuel (Kondamudi et al., 2008; Vardon et al., 2013; Lee et al., 2017; Zhao et al. 2019), surfactants (Deotale et al., 2019), pest control for coffee plants (Green et al. 2015), and other technologies (Le et al., 2019; Sun et al., 2019). However, we hope to have conveyed that coffee, be it green or brown, is deeply rooted in fundamental science; and the industry affords a playground of innovative solutions that take advantage of the unique structure and chemistry of the coffee bean material.

The following are a few actionable recommendations that will demonstrate the immediate impact of the science discussed herein.

Actionable Information

The following bullet points are simple experiments that will demonstrate some of the key scientific decisions made by professional baristas during the production of coffee beverages.

- Consider the bicarbonate concentration in the brew water. Try brewing with a high buffering water (e.g., Evian), and compare the flavor profile to that of a soft water (e.g., Dasani).
- Try to affect the size of the fine particulates by grinding coffee directly from the freezer. Cold coffee is known to produce more regular small particles, which should give rise to a cleaner cup.
- Control the brew ratio. Start by using 1–16 g of coffee, and adjust the brew ratio by either using more water or less coffee to arrive at a desired strength.
- Buy fresh, locally roasted beans. These products often focus on the expression of the production methods and celebrate the nuanced flavors that they afford.

References

Alonso-Salces, R. M., F. Serra, F. Reniero, and K. Héberger. "Botanical and Geographical Characterization of Green Coffee (*Coffea arabica* and *Coffea canephora*): Chemometric Evaluation of Phenolic and Methylxanthine Contents." *Journal of Agricultural and Food Chemistry* 57 (2009): 4224–4235.

Alves, R. C., I. M. C. Almeida, S. Casal, and M. B. P. P. Oliveira. "Isoflavones in Coffee: Influence of Species, Roast Degree, and Brewing Method." *Journal of Agricultural and Food Chemistry* 58 (2010): 3002–3007.

Amanpour, A., and S. Selli. "Differentiation of Volatile Profiles and Odor Activity Values of Turkish Coffee and French Press Coffee." *Journal of Food Processing and Preservation* 40 (2015): 1116–1124.

Ames, J. M. P. In *The Maillard Reaction: Consequences for the Chemical and Life Sciences*, edited by R. Ikan, ed. New York: John Wiley, 1996.

Andueza, S., M. A. Vila, M. Paz de Peña, and C. Cid. "Influence of Coffee/Water Ratio on the Final Quality of Espresso Coffee." *Journal of the Science of Food and Agriculture* 87 (2007): 586–592.

Anese, M., L. Manzocco, and M. C. Nicoli. "Modeling the Secondary Shelf Life of Ground Roasted Coffee." *Journal of Agricultural and Food Chemistry* 54 (2006): 5571–5576.

Arisseto, A. P., E. Vicente, M. S. Ueno, S. A. V. Tfouni, and M. C. D. F. Toledo. "Furan Levels in Coffee as Influenced by Species, Roast Degree, and Brewing Procedures." *Journal of Agricultural and Food Chemistry* 59 (2011): 3118–3124.

Avallone, S., J.-P. Guiraud, G. E. Olguin, and J.-M. Brillouet. "Polysaccharide Constituents of Coffee-Bean Mucilage." *Journal of Food Science* 65 (2000): 1308–1311.

Avallone, S., J.-P. Guiraud, G. E. Olguin, and J.-M. Brillouet. "Fate of Mucilage Cell Wall Polysaccharides during Coffee Fermentation." *Journal of Agricultural and Food Chemistry* 49 (2001a): 5556–5559.

Avallone, S., B. Guyot, J. M. Brillouet, E. Olguin, and J.-P. Guiraud. "Microbiological and Biochemical Study of Coffee Fermentation." *Current Microbiology* 42 (2001b): 252–256.

Bae, J., and K. H. Hong. "Optimized Dyeing Process for Enhancing the Functionalities of Spent Coffee Dyed Wool Fabrics Using a Facile Extraction Process." *Polymers* 11 (2019): 574.

Barbosa, M. S. G., M. D. S. Scholz, C. S. C. Kitzberger, and M. T. Benasso. "Correlation Between the Composition of Green Arabica Coffee Beans and the Sensory Quality of Coffee Brews." *Food Chemistry* 292 (2019): 275–280.

Batali, M. E., W. D. Ristenpart, and J.-X. Guinard. "Brew Temperature, at Fixed Brew Strength and Extraction, Has Little Impact on the Sensory Profile of Drip Brew Coffee." *Scientific Reports* 10 (2020): 16450.

Bhumiratana, N., K. Adhikari, and E. Chambers IV. "Evolution of Sensory Aroma Attributes from Coffee Beans to Brewed Coffee." *LWT—Food Science and Technology* 44 (2011): 2185–2192.

Bicchi, C. P., A. E. Binello, G. M. Pellegrino, and A. C. Vanni. "Characterization of Green and Roasted Coffees Through the Chlorogenic Acid Fraction by HPLC-UV and Principal Component Analysis." *Journal of Agricultural and Food Chemistry* 43 (1995): 1549–1555.

Blumberg, S., O. Frank, and T. Hofmann. "Quantitative Studies on the Influence of the Bean Roasting Parameters and Hot Water Percolation on the Concentrations of Bitter Compounds in Coffee Brew." *Journal of Agricultural and Food Chemistry* 58 (2010): 3720–3728.

Bradley, E. S., and C. H. Hendon. "The Impact of Solvent Relative Permittivity on the Dimerisation of Organic Molecules Well Below Their Solubility Limits: Examples from Brewed Coffee and Beyond." *Food & Function* 8 (2017): 1037–1042.

Bucheli, P., I. Meyer, A. Pittet, G. Vuataz, and R. Viani. "Industrial Storage of Green Robusta Coffee under Tropical Conditions and Its Impact on Raw Material Quality and Ochratoxin A Content." *Journal of Agricultural and Food Chemistry* 46 (1998): 4507–4511.

Buffo, R. A., and C. Cardello-Freire. "Coffee Flavour: An Overview." *Flavour and Fragrance Journal* 19 (2004): 99–104.

Calligaris, S., M. Munari, G. Arrighetti, and L. Barba. "Insights into the Physicochemical Properties of Coffee Oil." *European Journal of Lipid Science and Technology* 111 (2009): 1270–1277.

Cameron, M., D. Morisco, D. Hofstetter, E. Uman, J. Wilkinson, Z. C. Kennedy, S. A. Fontenot, W. T. Lee, C. H. Hendon, and J. M. Foster. "Systematically Improving Espresso: Insights from Mathematical Modeling and Experiment." *Matter* 2 (2020): 631–648.

Casal, S., M. B. Oliveira, and M. A. Ferreira. "HPLC/Diode-Array Applied to the Thermal Degradation of Trigonelline, Nicotinic Acid and Caffeine in Coffee." *Food Chemistry* 68 (2000): 481–485.

Colonna-Dashwood, M., and C. H. Hendon. *Water for Coffee.* Bristol, UK: Independent Publishing Network, 2015.

Cordoba, N., L. Pataquiva, C. Osorio, F. L. M. Moreno, and R. Y. Ruiz. "Effect of Grinding, Extraction Time and Type of Coffee on the Physicochemical and Flavour Characteristics of Cold Brew Coffee." *Scientific Reports* 9 (2019): 8440.

Crews, C., and L. Castle. "A Review of the Occurrence, Formation and Analysis of Furan in Heat-Processed Foods." *Trends in Food Science & Technology* 18 (2007): 365–372.

Daglia, M., A. Papetti, C. Aceti, B. Sordelli, V. Spini, and G. Gazzani. "Isolation and Determination of α-Dicarbonyl Compounds by RP-HPLC-DAD in Green and Roasted Coffee." *Journal of Agricultural and Food Chemistry* 55 (2007): 8877–8882.

DaMatta, F. M., and J. D. Cochicho Ramalho. "Impacts of Drought and Temperature Stress on Coffee Physiology and Production: A Review." *Brazilian Journal of Plant Physiology* 18 (2006): 55–81.

Davids, K. *Espresso: Ultimate Coffee.* 2nd ed. New York: St. Martin's Griffin, 2001.

De Moraes, M. H. P., and R. H. Luchese. "Ochratoxin A on Green Coffee: Influence of Harvest and Drying Processing Procedures." *Journal of Agricultural and Food Chemistry* 51 (2003): 5824–5828.

Deotale, S. M., S. Dutta, J. A. Moses, and C. Anandharamakrishnan. "Coffee Oil as a Natural Surfactant." *Food Chemistry* 295 (2019): 180–188.

Dias, E. C., R. G. F. A. Pereira, F. M. Borém, E. Mendes, R. R. de Lima, J. O. Fernandes, and S. Casal. "Biogenic Amine Profile in Unripe Arabica Coffee Beans Processed According to Dry and Wet Methods." *Journal of Agricultural and Food Chemistry* 60 (2012): 4120–4125.

Dias, R. C. E., E. F. F. Machado, A. Z. Mercadante, N. Bragagnolo, and M. T. Benassi. "Roasting Process Affects the Profile of Diterpenes in Coffee." *European Food Research and Technology* 239 (2014): 961–970.

Dold, S., C. Lindinger, E. Kolodziejczyk, P. Pollien, S. Ali, J. C. Germain, S. G. Perin, et al. "Influence of Foam Structure on the Release Kinetics of Volatiles from Espresso Coffee Prior to Consumption." *Journal of Agricultural and Food Chemistry* 59 (2011): 11196–11203.

Dong, W., L. Tan, J. Zhao, R. Hu, and M. Lu. "Characterization of Fatty Acid, Amino Acid and Volatile Compound Compositions and Bioactive Components of Seven Coffee (*Coffea robusta*) Cultivars Grown in Hainan Province, China." *Molecules* 9 (2015): 16687–16708.

Drioschi, B., Jr., R. C. Guarçoni, M. C. S. da Silva, T. G. R. Veloso, M. C. M. Kasuya, E. C. S. Oliveira, J. M. R. da Luz, T. R. Moreira, D. G. Debona, and L. L. Pereira. "Microbial Fermentation Affects Sensorial, Chemical, and Microbial Profile of Coffee under Carbonic Maceration." *Food Chemistry* 342 (2021): 128926.

Dsamou, M., O. Palicki, C. Septier, C. Chabanet, G. Lucchi, P. Ducoroy, M.-C. Chagnon, and M. Morzel. "Salivary Protein Profiles and Sensitivity to the Bitter Taste of Caffeine." *Chemical Senses* 37 (2012): 87–95.

Duarte, G. S., A. A. Pereira, and A. Farah. "Chlorogenic Acids and Other Relevant Compounds in Brazilian Coffees Processed by Semi-dry and Wet Post-harvesting Methods." *Food Chemistry* 118 (2010): 851–855.

Efthymiopoulos, I., P. Hellier, N. Ladommatos, A. Russo-Profili, A. Eveleigh, A. Aliev, A. Kay, and B. Mills-Lamptey. "Influence of Solvent Selection and Extraction Temperature on Yield and Composition of Lipids Extracted from Spent Coffee Grounds." *Industrial Crops and Products* 119 (2018): 49–56.

Ellero, M., and L. Navarini. "Mesoscopic Modelling and Simulation of Espresso Coffee Extraction." *Journal of Food Engineering* 263 (2019): 181–194.

Fabbri, A., C. Cevoli, L. Alessandrini, and S. Romani. "Numerical Modeling of Heat and Mass Transfer during Coffee Roasting Process." *Journal of Food Engineering* 105 (2011): 264–269.

Fadai, N. T., J. Melrose, C. P. Please, A. Schulman, and R. A. Van Gorder. "A Heat and Mass Transfer Study of Coffee Bean Roasting." *International Journal of Heat and Mass Transfer* 104 (2017): 787–799.

Fedele, V. Universal refractometer apparatus and method. US Patent 8,239,144. Filed March 31, 2010, issued August 7, 2012.

Ferrari, M., L. Navarini, L. Liggieri, F. Ravera, and S. F. Liverani. "Interfacial Properties of Coffee-Based Beverages." *Food Hydrocolloids* 21 (2007): 1374–1378.

Fox, G. P., A. Wu, L. Yiran, and L. Force. "Variation in Caffeine Concentration in Single Coffee Beans." *Journal of Agricultural and Food Chemistry* 61 (2013): 10772–10778.

Fretheim, I. "Water Activity in Specialty Green Coffee: A Long-Term Observational Study." Cafe Imports, February 11, 2019. https://www.cafe imports.com/north-america/blog/2019/02/11/water-activity-in-specia lty-green-coffee-a-long-term-observational-study-by-ian-fretheim/.

Frost, S. C., W. D. Ristenpart, and J.-X. Guinard. "Effect of Basket Geometry on the Sensory Quality and Consumer Acceptance of Drip Brewed Coffee." *Journal of Food Science* 84 (2019): 2297–2312.

Gloess, A. N., C. Yeretzian, R. Knochenmus, and M. Groessl. "On-line Analysis of Coffee Roasting with Ion Mobility Spectrometry–Mass Spectrometry (IMS–MS)." *International Journal of Mass Spectrometry* 424 (2018): 49–57.

Green, P. W. C., A. P. Davis, A. A. Cossé, and F. E. Vega. "Can Coffee Chemical Compounds and Insecticidal Plants Be Harnessed for Control of Major Coffee Pests?" *Journal of Agricultural and Food Chemistry* 63 (2015): 9427–9434.

Gu, B.-C. "Fermentation of Green Coffee Bean (*Coffea arabica*) with Lactic Acid Bacteria and Characterization of Its Biochemical Properties." Master's thesis, School of International Agricultural Technology, 2017.

Hendon, C. H., L. Colonna-Dashwood, and M. Colonna-Dashwood. "The Role of Dissolved Cations in Coffee Extraction." *Journal of Agricultural and Food Chemistry* 62 (2014): 4047–4050.

Hickinbotham, S. J. Method for producing wine by fall carbonic maceration. US Patent 4,615,887. 1986.

Houghton, R. W., and A. T. Kuhn. "Mass-Transport Problems and Some Design Concepts of Electrochemical Reactors." *Journal of Applied Electrochemistry* 4 (1974): 173–190.

Iaccheri, E., L. Ragni, C. Cevoli, S. Romani, M. D. Rosa, and P. Rocculi. "Glass Transition of Green and Roasted Coffee Investigated by Calorimetric and Dielectric Techniques." *Food Chemistry* 301 (2019): 125187.

Illy, E., and L. Navarini. "Neglected Food Bubbles: The Espresso Coffee Foam." *Food Biophysics* 6 (2011): 335–348.

Illy, A., and R. Viani. *Espresso Coffee: The Science of Quality.* 2nd ed. New York: Elsevier Academic Press, 2005.

Jousse, F., T. Jongen, W. Agterof, S. Russell, and P. Braat. "Simplified Kinetic Scheme of Flavor Formation by the Maillard Reaction." *Journal of Food Science* 67 (2002): 2534–2542.

Kang, D.-E., H.-U. Lee, M. Davaatseren, and M. S. Chung. "Comparison of Acrylamide and Furan Concentrations, Antioxidant Activities, and Volatile Profiles in Cold or Hot Brew Coffees." *Food Science and Biotechnology* 29 (2019): 141–148.

Kilcast, D., and S. Clegg. "Sensory Perception of Creaminess and Its Relationship with Food Structure." *Food Quality and Preference* 13 (2002): 609–623.

Kondamudi, N., S. K. Mohapatra, and M. Misra. "Spent Coffee Grounds as a Versatile Source of Green Energy." *Journal of Agricultural and Food Chemistry* 56 (2008): 11757–11760.

Król, K., M. Gantner, A. Tatarak, and E. Hallmann. "The Content of Polyphenols in Coffee Beans as Roasting, Origin and Storage Effect." *European Food Research and Technology* 246 (2020): 33–39.

Lankinen, M. A., A. Fauland, B.-I. Shimizu, J. Ågren, C. E. Wheelock, M. Laakso, U. Schwab, and J. Pihlajamäki. "Inflammatory Response to Dietary Linoleic Acid Depends on *FADS1* Genotype." *American Journal of Clinical Nutrition* 109 (2019): 165–175.

Lashermes, P., M. Combes, P. Trouslot, and A. Charrier. "Phylogenetic Relationships of Coffee-Tree Species (*Coffea* L.) as Inferred from ITS Sequences of Nuclear Ribosomal DNA." *Theoretical and Applied Genetics* 94 (1997): 947–955.

Le, V. T., T. M. Pham, V. D. Doan, O. E. Lebedeva, and H. T. Nguyen. "Removal of Pb(II) Ions from Aqueous Solution Using a Novel Composite Adsorbent of Fe_3O_4/PVA/Spent Coffee Grounds." *Separation Science and Technology* 54 (2019): 3070–3081.

Lee, D., Y.-G. Cho, H.-K. Song, S. J. Chun, S.-B. Park, D.-H. Choi, S.-Y. Lee, J. T. Yoo, and S.-Y. Lee. "Coffee-Driven Green Activation of Cellulose and Its Use for All-Paper Flexible Supercapacitors." *ACS Applied Materials & Interfaces* 9 (2017): 22568–22577.

Lee, L. W., M. W. Cheong, P. Curran, B. Yu, and S. Q. Liu. "Coffee Fermentation and Flavor—An Intricate and Delicate Relationship." *Food Chemistry* 185 (2015): 182–191.

Lockhart, E. E., C. L. Tucker, and M. C. Merritt. "The Effect of Water Impurities on the Flavor of Brewed Coffee." *Journal of Food Science* 20 (1955): 598–605.

López, J. A. S., M. Wellinger, A. N. Gloess, R. Zimmermann, and C. Yeretzian. "Extraction Kinetics of Coffee Aroma Compounds Using a Semi-automatic Machine: On-line Analysis by PTR-ToF-MS." *International Journal of Mass Spectrometry* 401 (2016): 22–30.

Ludwig, I. A., L. Sanchez, B. Caemmerer, L. W. Kroh, M. P. D. Peña, and C. Cid. "Extraction of Coffee Antioxidants: Impact of Brewing Time and Method." *Food Research International* 48 (2012): 57–64.

Lund, M. N., and C. A. Ray. "Control of Maillard Reactions in Foods: Strategies and Chemical Mechanisms." *Journal of Agricultural and Food Chemistry* 65 (2017): 4537–4552.

Maeztu, L., S. Andueza, C. Ibañez, M. P. de Peña, J. Bello, and C. Cid. "Multivariate Methods for Characterization and Classification of Espresso Coffees from Different Botanical Varieties and Types of Roast by Foam, Taste, and Mouthfeel." *Journal of Agricultural and Food Chemistry* 49 (2001): 4743–4747.

Manzocco, L., and M. C. Nicoli. "Modeling the Effect of Water Activity and Storage Temperature on Chemical Stability of Coffee Brews.' *Journal of Agricultural and Food Chemistry* 55 (2007): 6521–6526.

Mateus, M.-L., M. Rouvet, J.-C. Gumy, and R. Liardon. "Interactions of Water with Roasted and Ground Coffee in the Wetting Process Investigated by a Combination of Physical Determinations." *Journal of Agricultural and Food Chemistry* 55 (2007a): 2979–2984.

Mateus, M.-L., C. Lindinger, J.-C. Gumy, and R. Liardon. "Release Kinetics of Volatile Organic Compounds from Roasted and Ground Coffee: Online Measurements by PTR-MS and Mathematical Modeling." *Journal of Agricultural and Food Chemistry* 55 (2007b): 10117–10128.

McNutt, J., and Q. He. "Spent Coffee Grounds: A Review on Current Utilization." *Journal of Industrial and Engineering Chemistry* 71 (2019): 78–88.

Mintesnot, A., N. Dechassa, and A. Mohammed. "Association of Arabica Coffee Quality Attributes with Selected Soil Chemical Properties." *East African Journal of Sciences* 9 (2015): 73–84.

Moon, J.-K., H.-S. Yoo, and T. Shibamoto. "Role of Roasting Conditions in the Level of Chlorogenic Acid Content in Coffee Beans: Correlation with Coffee Acidity." *Journal of Agricultural and Food Chemistry* 57 (2009): 5365–5369.

Moreira, A. S. P., F. M. Nunes, M. R. M. Domingues, and M. A. Coimbra. "Galactomannans in Coffee." In *Coffee in Health and Disease Prevention*, edited by V. R. Preedy, 173–182. Cambridge, MA: Academic Press, 2015.

Müller, C., and T. Hofmann. "Quantitative Studies on the Formation of Phenol/2-Furfurylthiol Conjugates in Coffee Beverages toward the Understanding of the Molecular Mechanisms of Coffee Aroma Staling." *Journal of Agricultural and Food Chemistry* 55 (2007): 4095–4102.

Murthy, P. S., H. P. Sneha, K. Basavara, and K.-I. Kusumoto. "Modulation of Coffee Flavor Precursors by *Aspergillus oryzae* Serine Carboxypeptidases." *LWT—Food Science and Technology* 113 (2019): 108312.

Navarini, L., and D. Rivetti. "Water Quality for Espresso Coffee." *Food Chemistry* 122 (2010): 424–428.

Nicoli, M. C., A. L. Manzocco, and C. R. Lerici. "Antioxidant Properties of Coffee Brews in Relation to the Roasting Degree." *LWT—Food Science and Technology* 30 (1997): 292–297.

Nunes, F. M., M. A. Coimbra, A. C. Duarte, and I. Delgadillo. "Foamability, Foam Stability, and Chemical Composition of Espresso Coffee as Affected by the Degree of Roast." *Journal of Agricultural and Food Chemistry* 45 (1997): 3238–3243.

Ortiz, A., A. Ortiz, F. E. Vega, and F. Posada. "Volatile Composition of Coffee Berries at Different Stages of Ripeness and Their Possible Attraction to the Coffee Berry Borer *Hypothenemus hampei* (Coleoptera: Curculionidae)." *Journal of Agricultural and Food Chemistry* 52 (2004): 5914–5918.

Pangborn, R. M., I. M. Trabue, and A. C. Little. "Analysis of Coffee, Tea and Artificially Flavored Drinks Prepared from Mineralized Waters." *Journal of Food Science* 36 (1971): 255–362.

Parenti, A., L. Guerrini, P. Masella, S. Spinelli, L. Calamai, and P. Spugnolia. "Comparison of Espresso Coffee Brewing Techniques." *Journal of Food Engineering* 121 (2014): 112–117.

Perez-Locas, C., and V. A. Yaylayan. "Origin and Mechanistic Pathways of Formation of the Parent Furan—A Food Toxicant." *Journal of Agricultural and Food Chemistry* 52 (2004): 6830–6836.

Perez-Locas, C., and V. A. Yaylayan. "The Maillard Reaction and Food Quality Deterioration." In *Chemical Deterioration and Physical Instability of*

Food and Beverages, edited by L. H. Skibsted, J. Risbo, and M. L. Andersen, 70–94. Cambridge: Woodhead, 2010.

Pérez-Martínez, M., P. Sopelana, M. P. de Peña, and C. Cid. "Effects of Refrigeration and Oxygen on the Coffee Brew Composition." *European Food Research and Technology* 227 (2008): 1633–1640.

Prakash, I., S. R. Shankar, H. P. Sneha, P. Kumar, H. Om, K. Basavaraj, and P. S. Murthy. "Metabolomics and Volatile Fingerprint of Yeast Fermented Robusta Coffee: A Value Added Coffee." *LWT—Food Science and Technology* 154 (2022): 112717.

Rekik, F., H. van Es, N. Hernandex-Aguilera, and M. I. Gómez. "Linking Coffee to Soil: Can Soil Health Increase Coffee Cup Quality in Colombia?" *Soil Science* 184 (2019): 25–33.

Rivera, W., X. Velasco, C. Gálvez, C. Rincón, A. Rosales, and P. Arango. "Effect of the Roasting Process on Glass Transition and Phase Transition of Colombian Arabic Coffee Beans." *Procedia Food Science* 1 (2011): 385–390.

Roberts, D. D., P. Pollien, and C. Milo. "Solid-Phase Microextraction Method Development for Headspace Analysis of Volatile Flavor Compounds." *Journal of Agricultural and Food Chemistry* 48 (2000): 2430–2437.

Roos, Y. H. "Glass Transition Temperature and Its Relevance in Food Processing." *Annual Review of Food Science and Technology* 1 (2010): 469–496.

Ruosi, M. R., C. Cordero, C. Cagliero, P. Rubiolo, C. Bicchi, B. Sgorbini, and E. Liberto. "A Further Tool to Monitor the Coffee Roasting Process: Aroma Composition and Chemical Indices." *Journal of Agricultural and Food Chemistry* 60 (2012): 11283–11291.

Samsonowicz, M., E. Regulska, D. Karpowicz, and B. Leśniewska. "Antioxidant Properties of Coffee Substitutes Rich in Polyphenols and Minerals." *Food Chemistry* 278 (2019): 101–109.

Scheidig, C., M. Czerny, and P. Schieberle. "Changes in Key Odorants of Raw Coffee Beans during Storage under Defined Conditions." *Journal of Agricultural and Food Chemistry* 55 (2007): 5768–5775.

Scholz, B., and H. G. Maier. "Isomers of Quinic Acid and Quinide in Roasted Coffee." *Zeitschrift für Lebensmittel-Untersuchung Forschung* 190 (1943): 132–134.

Schroeder, T., P. B. da Silva, G. R. Basso, M. C. Franco, T. T. Maske, and M. S. Cenci. "Factors Affecting the Color Stability and Staining of Esthetic Restorations." *Odontology* 107 (2019): 507–512.

Silva, A. C. C., and F. C. Schmidt. "Vacuum Freezing of Coffee Extract under Different Process Conditions." *Food and Bioprocess Technology* 12 (2019): 1683–1695.

Smrke, S., M. Wellinger, T. Suzuki, F. Balsiger, S. E. W. Opitz, and C. Yeretzian. "Time-Resolved Gravimetric Method to Assess Degassing of Roasted Coffee." *Journal of Agricultural and Food Chemistry* 66 (2018): 5293–5300.

Solomon, T. "The Definition and Unit of Ionic Strength." *Journal of Chemical Education* 78 (2001): 1691.

Sopelana, P., M. Pérez-Martínez, I. López-Galilea, M. P. de Peña, and C. Cid. "Effect of Ultra High Temperature (UHT) Treatment on Coffee Brew Stability." *Food Research International* 50 (2013): 682–690.

Spencer, M., E. Sage, M. Velez, and J.-X. Guinard. "Using Single Free Sorting and Multivariate Exploratory Methods to Design a New Coffee Taster's Flavor Wheel." *Journal of Food Science* 81 (2016): S2997–S3005.

Spiro, M. "Modelling the Aqueous Extraction of Soluble Substances from Ground Roast Coffee." *Journal of the Science of Food and Agriculture* 61 (1993): 371–373.

Spiro, M., and R. M. Selwood. "The Kinetics and Mechanism of Caffeine Infusion from Coffee: The Effect of Particle Size." *Journal of the Science of Food and Agriculture* 35 (1984): 915–924.

Spiro, M., R. Toumi, and M. Kandiah. "The Kinetics and Mechanism of Caffeine Infusion from Coffee: The Hindrance Factor in Intra-bean Diffusion." *Journal of the Science of Food and Agriculture* 46 (1989): 349–356.

Stévart, T., G. Dauby, P. P. Lowry, A. Blach-Overgaard, V. Droissart, D. J. Harris, B. A. Mackinder, et al. "A Third of the Tropical African Flora Is Potentially Threatened with Extinction." *Science Advances* 5 (2019): eaav3473.

Sun, S., Q. Yu, M. Li, H. Zhao, and C. Wu. "Preparation of Coffee-Shell Activated Carbon and Its Application for Water Vapor Adsorption." *Renewable Energy* 142 (2019): 11–19.

Toledo, P. R. A. B., L. Pezza, H. R. Pezza, and A. T. Toci. "Relationship Between the Different Aspects Related to Coffee Quality and Their Volatile Compounds." *Comprehensive Reviews in Food Science and Food Safety* 15 (2016): 705–719.

Toschi, T. G., V. Cardenia, G. Bonaga, M. Mandrioli, and M. T. Rodriguez-Estrada. "Coffee Silverskin: Characterization, Possible Uses, and Safety Aspects." *Journal of Agricultural and Food Chemistry* 62 (2014): 10836–10844.

Uman, E., M. Colonna-Dashwood, L. Colonna-Dashwood, M. Perger, C. Klatt, S. Leighton, B. Miller, et al. "The Effect of Bean Origin and Temperature on Grinding Roasted Coffee." *Scientific Reports* 6 (2016): 24483.

van Boekel, M. A. J. S. "Kinetic Aspects of the Maillard Reaction: A Critical Review." *Molecular Nutrition and Food Research* 45 (2001): 150–159.

van Boekel, M. A. J. S. "Formation of Flavour Compounds in the Maillard Reaction." *Biotechnology Advances* 24 (2006): 230–233.

Vardon, D. R., B. R. Moser, W. Zheng, K. Witkin, R. L. Evangelista, T. J. Strathmann, K. Rajagopalan, and B. K. Sharma. "Complete Utilization of Spent Coffee Grounds to Produce Biodiesel, Bio-oil, and Biochar." *ACS Sustainable Chemistry & Engineering* 1 (2013): 1286–1294.

Vernon, G., and C. Parkanyi. "Mechanisms of Formation of Heterocyclic Compounds Maillard and Pyrolysis Reactions." In *The Chemistry of Heterocyclic Flavoring and Aroma Compounds*, edited by G. Vernin, 151–207. Chichester: Ellis Horwood, 1982.

Vilela, D. M., G. V. M. Pereira, C. F. Silva, L. R. Batista, and R. F. Schwan. "Molecular Ecology and Polyphasic Characterization of the Microbiota Associated with Semi-dry Processed Coffee (*Coffea arabica* L.)." *Food Microbiology* 27 (2010): 1128–1135.

Vincent, J. C. "Green Coffee Processing." In *Coffee*, edited by R. J. Clarke and R. Macrae, 1–33. New York: Elsevier, 1987.

Wang, X., L.-T. Lim, S. Tan, and Y. Fu. "Investigation of the Factors That Affect the Volume and Stability of Espresso Crema." *Food Research International* 116 (2019): 668–675.

Wang, X., W. William, Y. Fu, and L.-T. Lim. "Effects of Capsule Parameters on Coffee Extraction in Single-Serve Brewer." *Food Research International* 89 (2016): 797–805.

Wei, F., K. Furihata, M. Koda, F. Hu, T. Miyakawa, and M. Tanokura. "Roasting Process of Coffee Beans as Studied by Nuclear Magnetic Resonance: Time Course of Changes in Composition." *Journal of Agricultural and Food Chemistry* 60 (2012): 1005–1012.

Williamson, K., and E. Hatzakis. "Evaluating the Effect of Roasting on Coffee Lipids Using a Hybrid Targeted–Untargeted NMR Approach in Combination with MRI." *Food Chemistry* 299 (2019): 125039.

World Coffee Research. *Sensory Lexicon*. 2nd ed. College Station, TX: World Coffee Research, 2017.

Yadessa, A., J. Burkhardt, E. Bekele, K. Hundera, and H. Goldbach. "The Role of Soil Nutrient Ratios in Coffee Quality: Their Influence on Bean Size and Cup Quality in the Natural Coffee Forest Ecosystems of Ethiopia." *African Journal of Agricultural Research* 14 (2019): 2090–2103.

Yadessa, A., J. Burkhardt, E. Bekele, K. Hundera, and H. Goldbach. "The Major Factors Influencing Coffee Quality in Ethiopia: The Case of Wild Arabica Coffee (*Coffea arabica* L.) from Its Natural Habitat of Southwest and Southeast Afromontane Rainforests." *African Journal of Plant Science* 14 (2020): 213–230.

Zhang, S. J., F. De Bruyn, V. Pothakos, J. Torres, C. Falconi, C. Moccand, S. Weckx, and L. De Vuyst. "Following Coffee Production from Cherries to Cup: Microbiological and Metabolomic Analysis of Wet Processing of *Coffea arabica*." *Applied and Environmental Microbiology* 85 (2019): e02635-18.

Zhao, P., M. H. A. Shiraz, H. Zhu, Y. Liu, L. Tao, and J. Liu. "Hierarchically Porous Carbon from Waste Coffee Grounds for High-Performance Li–Se Batteries." *Electrochimica Acta* 325 (2019): 134931.

5

ARTISANAL CHEESE

MICHAEL H. TUNICK, SEANA DOUGHTY, AND MARÍA PATRICIA CHOMBO MORALES

Introduction

What is the difference between artisanal and mass-produced food? In many cases, the artisan will take extra steps to ensure a better and more distinctive product. This is especially true with fermented products such as cheese. France has the Appellation d'Origine Contrôlé (AOC) that specifies the traditional methods for making foods such as wine and cheese; the AOC describes artisanal cheese as being handmade by a dairy owner (though the milk can come from outside sources) and farmhouse cheese as being made by an independent producer using traditional methods and raw milk from a herd raised on the property (Masui and Yamada, 1996). Similarly, the American Cheese Society, which promotes and supports cheese made in the United States, states *"artisan* or *artisanal* implies that a cheese is produced primarily by hand, in small batches, with particular attention paid to the tradition of the cheesemaker's art, and thus using as little mechanization as

Michael H. Tunick, Seana Doughty, and María Patricia Chombo Morales, *Artisanal Cheese* In:
The Science and Craft of Artisanal Food. Edited by: Michael H. Tunick and Andrew L. Waterhouse,
Oxford University Press. © Oxford University Press 2023. DOI: 10.1093/oso/9780190936587.003.0006

possible in the production of the cheese. Artisan, or artisanal, cheeses may be made from all types of milk and may include various flavorings" (American Cheese Society, n.d.). They also define "farmstead cheese" as a product that "must be made with milk from the farmer's own herd, or flock, on the farm where the animals are raised" (American Cheese Society, n.d.). No US government agency has spelled out definitions of "artisanal" or "farmstead," so the society's definitions will suffice here.

The use of minimally modified milk and small-scale manufacturing techniques results in a wider range of flavors, textures, and varieties than mass-produced cheeses. Commercially made non-processed cheeses in the United States tend to have fewer and less intense flavors than those from small outfits. Industrial cheesemakers pasteurize milk from cows confined in barns, have computer-controlled operations, use large mechanical presses to quickly expel whey from the cheese curd, vacuum-package the product, and ship it out as quickly as practical to avoid tying up their storage facilities. In contrast, artisanal cheesemakers make their cheese by hand in small batches from cow, goat, sheep, and even water buffalo milk. They obtain a wider range of volatile flavor compounds by using milk from animals that feed on pasture plants. Some make cheese from raw milk, thus preserving indigenous microflora that would be inactivated by pasteurization. These bacteria cause the cheese to ripen faster and develop stronger flavors. Artisans use less expensive equipment, often age the cheese on wooden shelves, and keep it until they feel that the flavor has reached a peak.

Flavor Generation

Flavors in cheese arise from enzymatic breakdown of protein (mostly casein, with some whey proteins), fat and oils (known collectively as "lipids"), and carbohydrates (almost entirely lactose, with a small amount of citrate). Pasteurization partially or completely inactivates many enzymes indigenous to milk, so starter cultures are added (Grappin and Beuvier, 1997). These are mixtures of bacterial species and strains from *Enterococcus*, *Lactobacillus*, *Lactococcus*, *Leuconostoc*, *Streptococcus*, and other genera. Starter cultures are specifically selected for their ability to ferment lactose to lactic acid. As acid forms, the pH of the warm milk begins to drop. Rennet, which contains an enzyme that breaks one particular bond in one type of casein, is then added and causes the casein to coagulate, producing a solid curd and liquid whey. From then on, the cheesemaker removes whey by draining it from the vat and by applying pressure to the curd. The enzymes in the rennet and secreted by the starter culture remain active throughout processing, and as the starter bacteria die out during aging, the cells break open and release more enzymes (Tunick, 2007).

Many artisans use the same starter cultures as large-scale manufacturers, but those making raw milk cheese may find that the indigenous microflora in the milk can serve as a starter. Some use whey from a successful batch of cheese since the bacteria in the whey continue to grow. The enzymatic system of microflora in raw milk is much more complex than that of added starter bacteria, which strongly influences texture and flavor formation (Grappin and Beuvier, 1997; Ballesteros et al., 2006).

Carbohydrates

Since flavor is so important in cheese, we will give a summary of how it is formed. Lactic acid gives cheese an acidic taste but, more importantly, serves as a food source for the starter bacteria, which metabolize it to form characteristic cheese flavors. Carbon dioxide gas is also formed, which gives Swiss and other cheeses their eyes (holes). Citrate is also broken down into compounds that provide flavors that are reminiscent of cheese (McSweeney, 2004).

Protein

Proteins are degraded into smaller pieces known as "peptides," which are then broken down into amino acids, the building blocks of protein. Amino acids are responsible for basic tastes (bitter, sweet, sour, umami) and often form other compounds responsible for alcoholic, ammonia, freshly cut grass, pungent, and roasted flavors, among others (McSweeney, 2004).

Lipids

Most cheese flavors arise from the breakdown of triglycerides into fatty acids, which in turn break apart and form a slew of different compounds. Floral, fruity, goaty, mushroom, musty, and waxy flavors come from the degradation of lipids (McSweeney, 2004).

Often, fat levels are not adjusted in milk for artisanal cheese. Holsteins account for 90% of the milking cows in the United States and average 3.7% fat in their milk. Artisans often use other breeds for cheese milk, including Jerseys (4.6% fat) and Guernseys (4.5%).

Higher fat levels naturally lead to more lipid breakdown and more flavor compounds.

Manufacturing Process

Much artisanal cheese is made from raw milk, but when the milk is pasteurized, the heating is frequently carried out in a vat at 63°C (145°F) for 30 minutes. This vat pasteurization process is enough to kill off pathogens while not imparting a cooked flavor. Artisanal operations usually cannot afford a commercial pasteurizer, which treats the milk at 72°C (161°F) for 15 seconds. Such high-temperature, short-time pasteurization, which is always used in industry, inactivates indigenous milk enzymes, which eliminates some pathways for breaking down proteins and lipids.

Pressing of the curd is frequently done overnight under relatively low pressure, resulting in a texture that is not quite the same as that of mass-produced cheese. The larger presence of whey also increases the amount of ripening. Many artisanal producers do not vacuum-package cheese, instead using permeable wrapping that admits oxygen and lets bacteria survive longer.

Storage

Raw milk cheese in the United States must legally be aged at least 60 days at not less than 35°F to ensure that any pathogens die off, allowing further texture and flavor development to take place (Food and Drug Administration, 2017). Artisanal cheese may be stored in a "cheese cave" that is natural or dug from a hillside, which eliminates the costly need for temperature control and permits longer aging times.

Imported cheese may be shipped at an immature age, and the product may be stabilized to encourage shelf life. Artisanal cheese is produced locally, making preservatives and stabilizers unnecessary. Artisanal cheese is therefore ripened to full maturity for optimal flavor and texture.

Large-scale operations can maximize profits by replacing fat and protein with water up to the legal limit. Higher moisture content could lead to formation of off-flavors from more extensive enzymatic activity. These cheeses are stored for shorter times, which also results in decreased production of desirable flavors.

Most artisanal cheese is aged on wooden boards, which impart additional flavors. Wood provides good mechanical resistance and exceptional regulative water sorption properties that lead to a low-humidity loss of cheeses during ripening. Most cheese manufacturers believe that wooden shelves favor cheese rind development and improve the organoleptic qualities. Wooden shelves used for cheese ripening have been found to contain cellulose, hemicellulose, and lignin from wood; amides from whey protein; lipids from the rind; and proteins and polysaccharides from biofilms (Oulahal et al., 2009).

Variability

Another distinction between mass-produced and artisanal cheese is in the variability of the product. The components in milk are concentrated by a factor of roughly 10 in cheese, so an apparently minor difference in the milk may create a major difference later on. Artisanal producers take advantage of the fact that milk varies with feed, stage of lactation, and season. The food that the animals eat affects the resulting cheese, so grazing on different pastures will lead to changes in flavor. The fat and protein contents of bovine milk are at their lowest levels during the warmest months, decrease during the first 25–50 days of lactation, and increase thereafter (Urbach, 1990). Artisanal cheesemakers will often switch varieties or alter their procedures depending on the time of year. Large-scale companies in the United States try to have a uniform product at all times, so their milk suppliers have their cows on the same diet with staggered lactation periods. In Europe and other places, even mass-produced cheeses are made to "suit the season," and lactation periods are not placed on a schedule.

Another source of variability consists of bacteria that enter the process without being deliberately added. Many different species of microorganisms are present in the air in cheesemaking and storage facilities, which can affect the characteristics of the product. The action of these microbes may impart positive or negative flavor qualities.

Emotion

Consumers have cited non-sensory aspects of cheese and its consumption as being part of the sensory experience. Positive memories may become associated with an artisanal cheese that was consumed during a pleasant experience. Cheesemaker practices are also important to many people, who feel that small-scale production by families leads to a better product. Craftsmanship, moral sentiment, and "farm story" seem to influence the conceptions of consumers (Paxson, 2013; Lahne and Trubek, 2014).

Conclusions

None of the above is meant to imply that artisanal cheese is better for you than cheese produced on a large scale. The nutritional profiles are similar, though more research is needed on the health effects of components such as omega-3 fatty acids, which are more plentiful in milk from grazing animals. Many shoppers are happy to pay less for cheese made on an industrial scale and with knowing that they will get the same flavor and texture every time. Others are willing to pay more for artisanal cheeses because of their differing taste due to the greater diversity in the microflora. With a growing number of consumers who are more selective about their food choices, artisanal cheese cannot be surpassed.

Spotlight on Cotija

History

A good example of an artisanal cheese is cotija, one of the most consumed cheeses in Mexico. It has been reported that it has more than 400 years of history, and its origins are the hillsides of Michoacán and Jalisco, two states in the western part of Mexico, where the first settlements seeking gold and silver were established during the Spanish conquest. As time passed, a new mestizo culture arose, evidenced by husbandry practices and then by food. Hints of this culture remain today as the heritage of the region (Barragan and Chavez, 1998). Recently, a geographic indication (*indicación geográfica* [IG]), a Mexican commercial distinction similar to the French AOC, has been approved for this artisanal cheese. The IG specifies not only the process but also the geographic area that limits its manufacture, which today includes six municipalities that share the history, the culture, and the climate: Cotija, Los Reyes, and Tocumbo in Michoacán and Santa María del Oro, Quitupan, and Jiquilpan in Jalisco (González-Córdova et al., 2016). The IG aids artisans in marketing their cheese, with a better position due to this distinction. With the IG label consumers differentiate traditional cotija cheese from analogues and mass-produced versions of it (Pomeon, 2007).

Production

Today, the legacy of producing traditional cotija cheese is kept by nearly 200 families that live in the IG region. There, each family produces 180–250 L of milk for a 20- to 22-kg cheese wheel every day. The cheese production per season per family is an average of one to two tons. Traditional cotija is a cow's milk cheese, and it is only produced during the summer. Moreover, the IG stipulates that the herd must belong to the cheesemakers, be grass-fed during the whole season, and be Cebu as the dominant breed. This cheese is produced totally by hand, and if we consider the time when milking starts until a fresh cheese is transferred to the ripening shelves, processing takes 24 hours. Additionally, we would need to consider at least 3 months for careful ripening (González-Córdova et al., 2016). Since the activities to produce the cheese are distributed among all the members of the family, the manufacture of cotija cheese is often described as a job of patience by the artisans (see Figure 5.1).

Manufacturing Procedure

The manufacture starts with milking, which is always done by hand. This allows low counts of mastitis in the cow, although *Staphylococcus aureus* and coliforms sometimes are beyond specified limits in the milk. Basically, all the processing is performed at room temperature in a clean and fresh room in the

Figure 5.1 Hand-milking

family's house. Usually, time for transporting the milk and for renneting is shorter than 3 hours. Renneting is performed by adding bovine rennet to the milk. Coagulation takes place in 30–40 minutes. Afterward, the curd firmness is tested by marking a cross onto its surface with a wooden spoon, and when it has the desired firmness, it is blended, breaking the curd with the same item. The curd is allowed to set at the bottom of the vat, and most of the whey is drained. Afterward, the curd is formed into slabs that are stacked onto a stainless steel table (Figure 5.2). This stage allows the whey to drain until the curd develops a proper texture. When the curd stops draining, it is salted by hand. Afterward, it is molded using a wooden belt lined with a cactus- or cotton-fiber blanket (Figure 5.3). Once the cheese is molded, it is pressed for 15–20 hours. After pressing, fresh cheese is transferred to the ripening chamber, where the blanket is released but the belt is tightened over 15 days and then loosened (Figure 5.4). During ripening, cheese wheels are flipped over every other day, and the surface is rubbed with a clean cloth.

Characteristics

Traditional cotija is made with non-standardized (fat and protein levels are not adjusted) whole raw milk. Since it is not

Figure 5.2 Cutting the curd

heat-treated and no additives or starter are added, native microbial populations are responsible for fermentation and ripening.
At the end of 3 months of ripening cotija cheese reaches pH
5.4. During the ripening time the color of the rind turns golden
and its core ivory-creamy. It has a typical cylindrical shape,
weighs 20–22 kg, and has a dry and crumbly but creamy texture (Figure 5.5) (Van Hekken and Farkye, 2003). During the fermentation stage, several native lactic acid bacteria (LAB) grow
in the cheese core, and some of them remain viable throughout
ripening. A mixture of LAB, non-LAB, and yeasts develop
mostly on the surface.

Dynamics of the microbial populations and the specific
conditions of manufacture, including the hygiene practices,
the microbial quality of the milk, and the materials used during
processing, influence the acidity and the volatile profile of this
cheese (Flores-Magallon et al., 2011; Chombo-Morales et al.,
2016). In a recent work using the solid-phase microextraction/

Figure 5.3 Cotija press and a worker cleaning the belt used for molding the cheese

Figure 5.4 Small- and medium-sized ripening rooms

gas chromatographic–mass spectrometric technique, it was reported that the characteristic aroma of young traditional cotija comes from free fatty acids, terpenes, and methyl esters; after 90 days, ketones, alcohols, and other esters appear and their concentrations increase (Escobar-Zepeda et al., 2016). Some

Figure 5.5 The final product

of the compounds found in cotija cheese, such as butanoic, hexanoic, octanoic, and decanoic acids; their esters; and 2-heptanone, are associated with blue cheese and goat, fruity, musty, and rancid aromatic notes. These were reported in other cheeses such as cheddar or Parmesan with at least 6 months of ripening (Escobar-Zepeda et al., 2016). These results suggest that the biochemical changes in cotija cheese, such as lipolysis and oxidative reactions, take place faster during artisanal maturation than in other ripened cheeses (Montel et al., 2014). Since most of the Mexican cheeses are produced to be consumed fresh, the ripening stage of traditional cotija provides its distinctive characteristics.

References

American Cheese Society. "Cheese Definitions and Categories." n.d. https://www.cheesesociety.org/events-education/cheese-definitions/.

Ballesteros, C., J. M. Poveda, M. A. González-Viñas, and L. Cabezas. "Microbiological, Biochemical and Sensory Characteristics of Artisanal and Industrial Manchego Cheeses." *Food Control* 17 (2006): 249–255.

Barragan, L. E., and T. M. Chavez. "El queso Cotija se nos va de las manos." In *Manufacturas de Michoacán*, edited by V. Oikión Solano, 189–219. Michoacán, Mexico: El Colegio de Michoacán, 1998.

Chombo-Morales, P., M. Kirchmayr, A. Gschaedler, E. Lugo-Cervantes, and S. Villanueva-Rodríguez. "Effects of Controlling Ripening Conditions on the Dynamics of the Native Microbial Population of Mexican Artisanal Cotija Cheese Assessed by PCR-DGGE." *LWT—Food Science and Technology* 65 (2016): 1153–1161.

Escobar-Zepeda, A., A. Sanchez-Flores, and M. Quirasco Baruch. "Metagenomic Analysis of a Mexican Ripened Cheese Reveals a Unique Complex Microbiota." *Food Microbiology* 57 (2016): 116–127.

Flores-Magallon, R., A. A. Oliva-Hernandez, and J. A. Narvaez-Zapata. "Characterization of Microbial Trials Involved in the Elaboration of Cotija Cheese." *Food Science and Biotechnology* 20 (2011): 997–1003.

Food and Drug Administration. "Cheese and Related Cheese Products." 21 C.F.R., part 133 (2017).

González-Córdova, A. F., C. Yescas, A. M. Ortiz-Estrada, M. A. De la Rosa-Alcaraz, A. Hernández-Mendoza, and B. Vallejo-Cordoba. "Invited Review: Artisanal Mexican Cheeses." *Journal of Dairy Science* 99 (2016): 3250–3262.

Grappin, R., and E. Beuvier. "Possible Implications of Milk Pasteurization on the Manufacture and Sensory Quality of Ripened Cheese." *International Dairy Journal* 7 (1997): 751–761.

Lahne, J., and A. B. Trubek. "'A Little Information Excites Us': Consumer Sensory Experience of Vermont Artisan Cheese as Active Practice." *Appetite* 78 (2014): 129–138.

Masui, K., and T. Yamada. *French Cheeses*. New York: DK Publishing, 1996.

McSweeney, P. L. H. "Biochemistry of Cheese Ripening." *International Journal of Dairy Technology* 57 (2004): 127–144.

Montel, M. C., S. Buchin, A. Mallet, C. Delbes-Paus, D. A. Vuitton, N. Desmasures, and F. Berthier. "Traditional Cheeses: Rich and Diverse Microbiota with Associated Benefits." *International Journal of Food Microbiology* 177 (2014): 136–154.

Oulahal, N., I. Adt, C. Mariani, A. Carnet-Pantiez, E. Notz, and P. Degraeve. "Examination of Wooden Shelves Used in the Ripening of a Raw Milk Smear Cheese by FTIR Spectroscopy." *Food Control* 20 (2009): 658–663.

Paxson, H. *The Life of Cheese: Crafting Food and Value in America.* Berkeley: University of California Press, 2013.

Pomeon, T. *El queso Cotija.* San Salvador, México: FAO/IICA, 2007.

Tunick, M. H. "Origins of Cheese Flavor." In *Flavor of Dairy Products,* edited by K. R. Cadwallader, M. A. Drake, and R. J. McGorrin, 155–173. Washington, DC: ACS Books, 2007.

Urbach, G. "Effect of Feed on Flavor in Dairy Foods." *Journal of Dairy Science* 73 (1990): 3639–3650.

Van Hekken, D. L., and N. Y. Farkye. "Hispanic Cheeses: The Quest for Queso." *Food Technology* 57 (2003): 32–38.

6

THE CHEMISTRY
AND FLAVOR
OF ARTISANAL HONEY

KATIE UHL, ALYSON E. MITCHELL,
AND AMINA HARRIS

Introduction

Origins of Honey

Blossom honey is a naturally sweet substance produced by *Apis mellifera* (European honeybees) from the nectar of plants. European honeybees were introduced to North America in the 1600s and later to Australia, New Zealand, and parts of Asia (Crane, 1997). Other species of bees also produce honey, including *A. cerana, A. dorsata, A. laboriosa*, as well as several species of stingless bees including those from the genera *Melipona, Meliponula, Plebeia, Scaptotrigona, Tetragonisca,* and *Tetragonula* (Crane, 1997; Thrasyvoulou et al., 2018; Souza et al., 2006). Honeydew is a distinct type of honey produced by honeybees foraging on the sweet excretions of scale insects,

Katie Uhl, Alyson E. Mitchell, and Amina Harris, *The Chemistry and Flavor of Artisanal Honey* In: *The Science and Craft of Artisanal Food.* Edited by: Michael H. Tunick and Andrew L. Waterhouse, Oxford University Press.
© Oxford University Press 2023. DOI: 10.1093/oso/9780190936587.003.0007

especially aphids, that feed on tree sap (Crane, 1999). Honeybees collect and process this material the same as floral nectar, and the *Codex Alimentarius* (Codex Alimentarius Commission, 2001) includes honeydew honey in its honey standards.

Honey is an ancient food, and evidence for human honey hunting dates as far back as 8000–6000 BCE (Crane, 1997). The earliest documentation of honey hunting is found in a Mesolithic rock drawing in the Cuevas de la Araña of Spain (Figure 6.1) (Crane,

Figure 6.1 Woman gathering honey depicted in an 8,000-year-old rock painting in the Cuevas de la Araña, near Bicorp, Spain.

Francisco Hernandez Pacheco, 1924; Museo nacionale de ciencas naturales, Madrid, Spain.

1997). The oldest evidence of actual beekeeping (i.e., providing bees with an artificial space to build a comb) comes from Egypt and dates back to 2450 BCE (Kristy, 2017). In the ancient world, honey and date fruit were the principal sweeteners of food. As beekeeping or apiculture became established, larger amounts of honey became available; honey increasingly was used not only as a sugar source but as a medicinal ingredient and in religious ceremonies, and it became important in bartering and for trade (Cajka et al., 2009). Ancient documents, such as the Ebers Papyrus, describe the use of honey to treat a wide range of illnesses including coughs, burns, wounds, digestive disorders, and even blindness. Cultures around the globe have documented the use of honey in traditional medicines for the treatment of a variety of diseases, in particular the healing of wounds and intestinal diseases. More re-cent research has shown that honey exhibits strong antibacterial activity, wound-healing properties, and potent antioxidant and anti-inflammatory activity (Majtan, 2014; Yaghoobi et al., 2013; Wahdan, 1998).

Beekeeping changed dramatically in the mid-1800s with the de-velopment of the Langstroth hive (Figure 6.2). The Langstroth hive (patented in 1852) was the first practical movable-frame hive with vertically hung frames in which bees would build honeycomb. The Langstroth hive was followed by the invention of various honey extractors (e.g., hand-crank centrifuge, sieve) which are still in use today, though many have been mechanized or enlarged for com-mercial applications.

Bees make honey by collecting nectar and pollen from plants and transporting it to the hive. The nectar is converted into honey, and the pollen is used as a source of protein and lipids for the hive

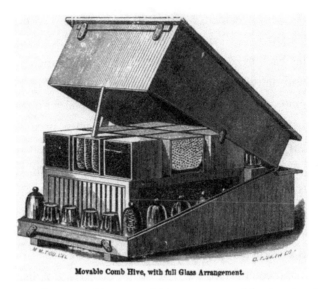

Movable Comb Hive, with full Glass Arrangement.

Figure 6.2 The original Langstroth hive, courtesy of the *American Bee Journal*

(Crailsheim, 1992). The collected nectar is stored in a specialized honey stomach (the crop) of the bee. The pollen attaches to tiny hairs on the bee's body. Floral nectars are composed primarily of water (~80%), sucrose, glucose, and fructose. During the flight back to the hive, the nectar mixes with enzymes excreted by salivary glands in the honey stomach. Thus begins the process of turning the nectar into honey via the inversion of sucrose. When the honey stomach is full, the forager bee will return to the hive and transfer the nectar to a younger worker bee. Bees will pass the nectar from one bee to another, through a process of repeated regurgitation, further inverting sucrose into glucose and fructose. The bees then deposit the regurgitated honey into unsealed honeycomb cells. At this point, the nectar still has a high water content (~50%). Bees fan the fresh nectar with their wings to evaporate the water from the nectar and reduce the water content to ~17%, which helps prevent fermentation of the honey. The bees then cap the honeycomb cells with wax, and the "ripened" honey is stored as a food source for the beehive.

Types of Honey

In general, bees are economical in their foraging habits and forage from within a fraction of a kilometer to several kilometers of the beehive (Ratnieks and Shackleton, 2015). Honey can therefore be classified based upon the floral source from which the bees gather nectar. *Monofloral* honey is made primarily from a specific type of flower nectar (e.g., orange blossom, clover) and is often preferred by consumers because it has distinctive flavors and aromas associated with the nectar source from which it is derived. Some of the most valued monofloral honeys include manuka (Australia, New Zealand), thyme (New Zealand, Greece), and tupelo (United States). In the United States, there are more than 300 different types of monofloral honey available, such as alfalfa, avocado, buckwheat, clover, eucalyptus, white kiawe, orange blossom, sage, sourwood, and star thistle.

Multifloral honey, such as wildflower, is made by bees foraging on the nectar of many types of flowers. The taste of wildflower honeys can vary widely. The flavor of these honeys is usually less distinct than that of monofloral honeys. Artisanal honeys are usually monofloral honeys, while commercially processed honeys usually consist of a mixture of two or more honeys differing in floral source, color, flavor, or geographic origin.

Major Components

Although honey is a highly variable natural product whose composition is influenced by many factors, including honeybee species, forage source, hive conditions, beekeeping practices, and environmental and geographical conditions, the major components of honey are carbohydrates (~80% dry weight) and water (~17%) (Crane, 1999; White, 1978). The monosaccharides fructose and glucose represent ~75% of the carbohydrates found in honey, whereas a range of disaccharides (e.g., sucrose, maltose, trehalose) as well as tri- and oligosaccharides represent about 10%–15% of the carbohydrates (Bogdanov et al., 2008). Additionally, honey contains a variety of minor components that include flavonoids (Pyrzynska and Biesaga, 2009; Truchado et al., 2009), phenolics (Pyrzynska and Biesaga, 2009), glycosides (Truchado et al., 2009), protein (~0.5% mainly as enzymes), and free amino acids (Bogdanov et al., 2008; Iglesias et al., 2006).

Sugar composition depends mainly on the honey's botanical and geographical origins, and it is affected by processing and storage (Escuredo et al., 2013; Tornuk et al., 2013). The concentrations of fructose and glucose, as well as the ratio between them, are useful indicators for the classification of monofloral honeys (Kaškoniene et al., 2010). In almost all types of honey, fructose (30%–40%) is in greatest proportion relative to glucose (25%–40%). Exceptions include honey from rape (*Brassica napus*) and dandelion (*Taraxacum officinale*) (Escuredo et al., 2013).

Honey Quality Characteristics

Color, texture, aroma, and flavor are key quality characteristics of blossom honey. The variety of plants from which nectars are collected has the greatest influence on the chemical composition of the honey. Honey color ranges from nearly colorless to an almost black amber tone, and its flavor varies from mild to strong, depending on floral nectar source, mineral content, temperature at which the honey remains in the hive, and storage time (Gámbaro et al., 2007). In the United States, the color of honey is graded based upon US Department of Agriculture (USDA)–approved color standards (Table 6.1) and includes colors that range from "water white" to "dark amber" (USDA, 1985). Organic acids can also affect the color and flavor of honey, and honey acidity is shown to decrease with increased ash content (Finola et al., 2007). In general, honey is an acidic product with a pH that can range from 3.33 to 6.56 depending on forage source and honeybee

Table 6.1 Color Designations of Honey

USDA color standards designation	Color range Pfund scales (mm)	Optical density*
Water white	≤8	0.0945
Extra white	>8–17	0.189
White	>17–34	0.378
Extra light amber	>34–50	0.595
Light amber	>50–85	1.389
Amber	>85–114	3.008
Dark amber	>114	>3.008

*Optical density (absorbance) = log (100/percent transmittance) at 560 nm for 3.15 cm thickness for caramel—10 glycerin solutions measured versus an equal cell containing glycerin.

species (White, 1978; Biluca et al., 2016). Honey composition and color are also affected by post-harvest thermal processing, packaging, and storage times due to fermentation and oxidation (Moreira et al., 2010), which is discussed further below.

Processed honey sold in the United States is given a quality grade (i.e., US Grade A, B, C, or substandard) based upon flavor and aroma, clarity, and absence of defects (USDA, 1985). Artisanal honey producers and consumers tend to favor raw or unpasteurized honey as the flavor profiles of honeys derived from different floral nectar sources are more distinct in these honeys as compared with processed honey.

Crystallization of Honey

Honey is a supersaturated sugar solution that will spontaneously crystallize. The relative ratios of fructose to glucose and glucose to water relate directly to the rate of crystallization of honey. Because glucose has lower water solubility than fructose, honeys with a higher ratio of glucose to fructose will crystalize more rapidly (e.g., rape and dandelion) than those with a higher fructose-to-glucose ratio (e.g., sage and tupelo). The higher the glucose level and the lower the water content of honey, the faster the crystallization process will occur. Some honeys may crystallize within a few weeks after extraction from the comb, whereas others may take years. The more rapidly honey crystallizes, the smaller the crystals will be and the finer the texture of the honey. Honey with a higher water content may crystallize unevenly and separate into crystallized and liquid layers. The speed of honey crystallization depends on its composition and the presence of nuclei for crystallization, such as pollen grains and pieces of beeswax in the honey. In general, unheated and unfiltered honey contains bits of wax, pollen, and propolis, which lead to faster crystallization. Honey prepared for the commercial market is usually heated and filtered, dissolving any sugar crystals and removing foreign particles that slow the crystallization process.

Storage temperature can affect the crystallization of honey. Honey crystallization is most rapid at 10–15°C and slows at temperatures below 10°C. The lower temperatures increase the viscosity of honey and inhibit the diffusion and formation of glucose crystals. Honey resists crystallization at temperatures greater than 25°C. Glucose crystals will dissolve in honey heated above 40°C; however, these temperatures may damage honey due to Maillard reactions.

Understanding Sources of Chemical Differences in Honey

Thermal Processing

Consumers expect a liquid, non-crystallized honey, but raw honey can crystallize or, though more unlikely, ferment faster than what consumers want. Therefore, industrial production of honey includes a multistage thermal processing, including low-temperature heating (about 45–55°C) for 4–5 days to liquefy the honey and melt any sugar crystals (Escriche et al., 2014; Wang et al., 2004). This is followed by heating at a higher temperature (about 80–85°C) for several minutes to pasteurize the honey in order to destroy yeast or other microorganisms that could ferment it.

Thermal processing can affect several factors of quality in the honey, including formation of Maillard reaction products including hydroxymethylfurfural (HMF). HMF is the main furan measured in honey as an indicator of heat processing history and freshness. This compound is created by the degradation of sugars in combination with certain amino acids and is accelerated with heat. HMF levels can be determined by reverse-phase high-performed liquid chromatography (HPLC) with detection using ultraviolet (UV) light or UV-visible spectrophotometry with detection at 285 nm (Pita-Calvo et al., 2017). The European Union and the *Codex Alimentarius* has set a limit of 40 mg/kg of HMF in honey with an exception of 80 mg/kg for honeys from tropical regions (Da Silva et al., 2016). Fresh raw honey typically has values below 10 mg/kg. With the heat treatments applied for industrial-scale honey production, the values of HMF can significantly increase, sometimes by more than 50%. HMF levels also tend to increase over time and can significantly increase within just 6 months of storage (Wang and Li, 2011). While one study demonstrated that the levels of HMF in heat-treated and stored honey were still well below the EU legal limit (Escriche et al., 2009), another revealed that HMF levels easily can exceed the 40-mg/kg limit in several days when exposed to temperatures above 35°C (Escriche et al., 2008). These studies have also shown that different types of honey respond differently to thermal processing, making it more difficult to estimate the impacts of heat and/or storage on the levels of HMF for any given honey.

The volatile profile of honeys can also be influenced by thermal processing. Volatile organic compounds (VOCs) include the aroma compounds distinct to varietal honeys and can be used to characterize honey varietals for botanical source and geographic region of production, especially when pollen analysis is not possible. Studies have found over 400 volatile compounds in honeys, including alcohols, esters, aldehydes, ketones, furans, hydrocarbons, phenols, and nitrogen-containing compounds (Wang and Li, 2011). VOCs are measured by gas chromatography with mass spectrometry (GC/MS). There are several methods to capture the volatiles in honey, including dynamic headspace (HS) analysis, purge and trap, and solid-phase microextraction (SPME) (Wang and Li, 2011). Studies that have investigated the effect of industrial thermal processing on the volatile profile of honeys have found that some groups of volatiles, namely alcohols and aldehydes, are more impacted than others (Escriche et al., 2009). The concentration of certain volatile compounds tends to increase with thermal processing, especially furan compounds, due to degradation of sugars and proteins in honey during heating. There may be perceived organoleptic differences in honeys that undergo heat treatments due to the change in volatile profiles; however, honeys of different botanical origins respond differently to thermal processing, making it challenging to predict the effects of thermal processing on a given honey.

Thermal processing can also reduce the concentration of important compounds that enzymes create in honey. The enzyme content of honey is usually determined by measuring the activity of diastase (the group of α-, β-, and γ-amylases), invertase, and glucose oxidase. Diastase converts starch to short-chain sugars, and the diastase activity is reported as the diastase number (DN) (Subramanian et al., 2007). The minimum DN for processed honey is 8, according to the European Regional Standard for honey. The DN tends to decrease with thermal processing as diatase is heat-labile and can be significantly decreased after pasteurization (Escriche et al., 2014; Subramanian et al., 2007). Glucose oxidase breaks down glucose to form both gluconic acid, the primary organic acid in honey, and hydrogen peroxide. It has been shown that both of these compounds can be dramatically reduced by thermal processing, which can impact the flavor and the health-promoting properties of the honey (Wang et al., 2004). Invertase is responsible for hydrolyzing sucrose, but currently there is not strong research into the effects of thermal processing on this enzyme.

Raw honey naturally contains yeast, which can lead to the fermentation of honey if the water activity is high enough. Fermentation can lead to the formation of desirable products (like honey ferments) and even mead. However, fermentation is considered an undesirable trait in most commercial honeys, so a higher heat pasteurization step is used to destroy yeast microorganisms, reduce the moisture content, and prevent fermentation (Subramanian et al., 2007). While the moisture level in raw honey is usually too low for fermentation to occur, it can contain about 1×10^5 colony forming units (cfu)/g of yeast, and thermal processing can reduce this to about 500 cfu/g, the commercially acceptable level (Subramanian et al., 2007). Alternatives to conventional heating for honey include microwave heating, infrared heating, ultrasound processing to liquefy crystals, and membrane ultrafiltration to non-thermally remove yeasts and other microorganisms that could possibly ferment the honey if the moisture level was too high (Subramanian et al., 2007).

Sensory Analysis of Honey

Sensory analysis is a helpful tool for honey characterization, ranging from classification of botanical source, assessing quality, and identification defects. Sensory analyses can also be powerful when used in conjunction with physiochemical and pollen analysis (Marcazzan et al., 2018; Piana et al., 2004). The European Commission stipulates that honey cannot contain any foreign taste and cannot have begun the fermentation process. The Italian Register of Experts in the Sensory Analysis of Honey has developed standard terminology and tasting methods, in addition to the International Honey Commission of the Apimondia (Piana et al., 2004). One of the efforts to unify descriptors for the sensory analysis of honey has been the creation of honey flavor wheels, such as the one produced in 2014 by the University of California at Davis Honey and Pollination Center seen in Figure 6.3. This honey flavor wheel was developed used 20 trained panelists to generate over 100 descriptors in groups such as animal or fruity and then further divides into secondary and sometimes tertiary terms, such as leather in the "woody" family or lime under citrus fruit in the "fruity" family.

Two methods of descriptive analysis are the semi-quantitative method and the profile method. A descriptive semi-quantitative method can be used with all honey types, whereas a profile method is better used when analyzing honey of a single botanical origin (Marcazzan et al., 2018). There are now guidelines for selecting panelists, conducting trainings, preparing samples, conducting the sessions, and tasting the honey samples (Marcazzan et al., 2018; Pinan et al., 2004). While this standardized method has helped advance the field of honey sensory analysis, there is still more progress to be made. There is a need for reference monofloral

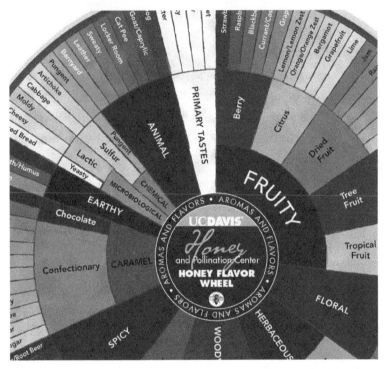

Figure 6.3 A partial representation of the Honey Flavor Wheel developed by the UC Davis Honey and Pollination Center

honeys, free of defects and from reliable beekeepers, in order to establish standards and train panelists.

Sensory analysis has been performed on a variety of monofloral, artisanal honeys already. Chestnut honeys from three regions in Spain were analyzed by a sensory panel in addition to GC/MS analysis (Castro-Vázquez et al., 2010). The data from the descriptive analysis panel correlated with the GC/MS data. For example, high levels of linalool, hotrienol, and epoxylinalool led to significantly higher "floral" and "fruity" odors from northwest Spain, whereas the "woody" and "toasty caramel" aromas from γ-butyrolactone and oak lactone were distinctive for honeys from northeast Spain.

In another study, quantitative descriptive analysis was performed on 55 artisanal honeys from Madrid, Spain, which were divided into three classes: floral, blend, and honeydew (González et al., 2010). Aroma attributes measured by the sensory panel included fruity, flowery, warm, aromatic, vegetal, animal, spoiled, and chemical. Other attributes included aroma intensity, aroma persistence, sweetness, bitterness, acidity, color, granularity, adhesiveness, and viscosity. The sensory panel did find blended honeys to be significantly more bitter than floral honeys, and floral honeys had significantly lower aroma persistence than blended and honeydew honeys.

A study on Austrian and Croatian monofloral honeys conducted volatile and sensory analyses to better understand the flavor components of dandelion, fir tree, linden tree, chestnut tree, robinia, orange, lavender, and rape honeys (Siegmund et al., 2018). HS/SPME coupled with GC/MS was used to analyze volatile compounds. The results showed that lavender honey had higher levels of phenyl acetaldehyde and β-damascenone, contributing to a floral and honey-like aroma. Orange honey had significantly higher methyl anthranilate, a marker of citrus honey. This compound, along with lilac aldehydes, contributed to the unique aroma and flavor of orange honey. Furthermore, the results indicated the use of essential oils in several honey samples, showing that these methods of analysis can be used to assess quality and identify defects.

A more recent study at the University of California at Davis in 2017 evaluated clover, orange blossom, and buckwheat honeys in order to determine if the varieties could be distinguished by their volatiles, in conjunction with sensory and pollen analyses, in addition to sugar, amino acid, and mineral analyses via nuclear magnetic resonance (NMR). Five samples of each variety were analyzed by HS-SPME–GC/MS, and data were evaluated using principal component analysis (PCA) to identify correlations. The levels of volatile compounds identified were compared within and between the varieties. Compounds were identified using seven ions per peak and matching MS spectra against a library using >90% confidence. This stringent identification eliminates noise and misidentified compounds. Using these criteria, 125 volatile compounds were identified in the samples, some of which were used to distinguish the varietals. PCA demonstrated differentiation of honeys based upon their floral source (Figure 6.4). For example, the clover honey samples had significantly higher cinnamaldehyde, coumarin,

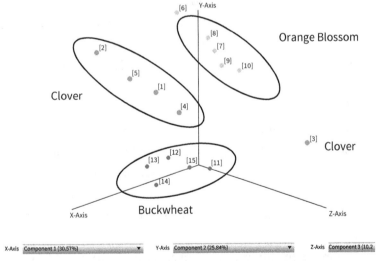

Figure 6.4 Principal component analysis distinguishing clover, orange blossom, and buckwheat honey by headspace solid-phase microextraction coupled with gas chromatography–mass spectrometry

3-hexanol, 2-hydroxybenzaldehyde, phenethyl alcohol, and benzaldehyde than the other honeys. However, one clover honey sample was unlike all other honey samples, as seen on the right side of the PCA in Figure 6.4. This sample did not match the other clover honey samples, having significantly higher *trans*-linalool oxide (furanoid), decanal, isophorone, benzeneacetaldehyde, eugenol, methyl nonanoate, 2-methoxy-phenol, and acetic acid. It is believed that this honey was a mixture of other honey types, giving it a unique volatile profile unlike the other clover honeys, and demonstrating the power of PCA in the identification of honey varietals. This honey was also an outlier in the sensory panel and NMR analysis and demonstrates the value of using HS-SPME–GC/MS volatile profiles and PCA to distinguish honey varietals.

The orange blossom honeys were distinguished by their levels of limonene, lilac aldehydes, linalool, methyl anthranilate, β-myrcene, 3-methyl-benzoic acid, thymol, and 2-heptanol. These aroma-active volatile compounds align with the citrusy notes determined by the sensory panel. The buckwheat honeys were determined to have strong animal and microbiological attributes by the trained sensory panel, which was supported by the distinguishing volatiles, including 5-methyl-2-furancarboxaldehyde, butanoic acid, 3-methylbutanoic acid, 2-furnamethanol, *endo*-borneol, furfural, hexanoic acid, p-cresol, and 2-methoxymethyl-furan.

Adulteration

Honey adulteration and counterfeiting is a common problem in the global food marketplace, resulting in significant economic fraud. Artisanal honeys are a target of adulteration because they are a high-price item with high demand. Adulteration techniques include dilution of the honey with water and extension with sugar-based syrups, such as beet syrup, corn syrup, high-fructose corn syrup, invert sugar, cane syrup, and dextrose syrup (Pita-Calvo et al., 2017; Kelly et al., 2006). Techniques to identify adulteration include NMR, HPLC, GC, stable carbon isotope ratio analysis, Raman spectroscopy, near-infrared spectroscopy, and mid-infrared spectroscopy. Honey parameters that can be measured to identify adulterated product include sucrose, the glucose–fructose–sucrose ratio, proline, and HMF (Pita-Calvo et al., 2017). For example, if the proline concentration is <180 mg/kg, the honey is considered to be either adulterated or nonripe. Sucrose has been a target parameter for adulteration. Authentic honey contains only about 5% sucrose, though levels can change depending on timing of harvest or maturity of the honey. The oligosaccharide profile can be used to look at honey further to identify adulteration with a syrup as honey and invert sugar have different oligosaccharide profiles. Similarly, a high level of HMF can occur due to adulteration with invert sugar because the acid used for hydrolysis of sucrose can produce HMF, though it may remain below the legal limit and requires more thorough testing to diagnose (Pita-Calvo et al., 2017; Wang and Li, 2011).

The tools used for botanical authentication are also important for identifying honeys that have been adulterated or mislabeled. This is a complex challenge because the volatile profile of honey can be altered by many variables, like botanical source, geographic region, harvest year, and storage conditions, so finding evidence of adulteration among all the factors can be difficult. There is a need for honey standards, known samples from trusted sources that can be used to establish volatile and sensory profiles and limits for other parameters.

Artisanal Honey Production

Here, we highlight the story of an artisanal packer, Z Specialty Food, LLC, from the perspective of Amina Harris, who discusses how artisanal packers differ from commercial packers.

Along with my son and daughter, I own an artisanal honey packing company, Z Specialty Food, LLC. We have been packing varietal honeys for over 40 years. As a small packer, and once a tiny packer, I am intimately aware of how we process and package honey and how this process significantly differs from commercial packers. Many of the differences are in the details.

We began, like many other packers and beekeepers large and small and tiny, because our founder, Ishai Zeldner, seen in Figure 6.5, loved honey and honeybees. He wanted to share his enthusiasm and passion. Though Z Specialty has grown from a small outbuilding of about 150 square feet to a 20,000–square foot facility, we still share the excitement about honey and bees that Ishai had when he filled the first bottles with Yellow Star Thistle Honey collected from beehives in the fields of northern California.

Z Specialty Food (called Moon Shine Trading Company when it began in 1979) keeps the same values it held at its beginnings. We pack in small quantities, and we often buy our honey in small quantities of a couple barrels up to 20 or so. We personally know our beekeepers. We are looking for only the very best of each of the varietals we carry, so a strong relationship between us, the packer, and the beekeeper is key to our final, bottled product. We strive to ensure that the honey is warmed gently, settled, and strained. Our process ensures that the naturally occurring pollen and enzymes are present in the bottled product. Our honeys will crystallize, and we are proud of that!

Figure 6.5 The Zeldner family selling honey at the children's pre-school about 1991.
Photo courtesy of Amina Harris, Z Specialty Food, LLC.

Over the years, we learned more and more about each of our varietals. We became more selective, searching for the flavors and aromas that would let us know we had pure varietal honeys. Z Specialty is pleased that most of its varietal honeys are about 85% or more naturally true to their source flowers.

Varietal honey selection is a key area in which the differences between the artisanal producer and the commercial packer vary. The commercial packer looks for a range of characteristics in honeys and then blends them to a specific flavor profile that will rarely crystallize. The artisan packer prefers to purchase the best of each varietal available and offers that natural profile to the consumer.

This is a growing issue for both the artisanal packer and the consumer. For most of historical time, honey has been collected, and consumers enjoyed the honey indigenous to their area. People were thrilled with something so sweet and wonderful. They didn't care if it was floral or spicy, dark or light in color. It was honey, and they were happy. It was the honey from their region, and they used it.

Eventually both beekeepers and consumers worldwide began to understand that each local honey was unique. Creamy lavender honey came from the fields of Provence. Aromatic orange blossom honey came from the orchards of Florida and California. By the 1960s, beekeepers would seek out specific forage areas to collect unique varietals. Many consumers—honey aficionados—would collect these regional varietal honeys. This is the population to whom Z Specialty Food sells honey.

As these honeys gained regional significance, the field of sensory science was being developed by Rose Marie Pangborn at the University of California at Davis during the 1960s and 1970s. Pangborn was a pioneer in the field, eventually coauthoring *Principles of Sensory Evaluation* (Amerine et al., 1965), which became the primary text for sensory course work and research for over 20 years.

Though much of this research has been applied to creating guidelines for the evaluation of wine, olive oil, and coffee, only the Italian Registry of Experts in the Sensory Analysis of Honey has applied the research to honey. In so doing, they have created sensory criteria for the evaluation of about 20 honey varietals found throughout Italy.

Why hasn't the same been done in the United States? There are approximately 300 varietal honeys being collected throughout the country. There has been little quantitative research done to describe these honeys and explain their differences. At the same time, consumer interest has grown, and more artisanal packers strive to meet the interest and the demand. In the world of honey packers, this is seen as a strong income opportunity. Commercial packers have added boutique divisions to their company trying to capitalize on that revenue. Without guidance and regulation, it will be difficult to prove which pure varietal honey might be in the jar.

This is problematic for the small packer who hopes to offer single pure varietals—there are no scientific criteria in place or any government directive to help them state the purity of their product. A honey that is pure orange blossom can appear on the grocery store shelf next to a blended orange blossom, both labeled as "pure orange blossom." Furthermore, consumers don't have the education to know what each varietal honey should really taste like.

A company like Z Specialty Food needs to have a guideline of criteria to select for great varietal honey. Essentially, we have done that ourselves since we began. However, times are changing. In the beginning, we had to rely on each beekeeper who would tell us about hive placement and how they felt the honey should taste and did taste. We learned the characteristics of each varietal we carried by trial and an occasional error. Our company established our own profile for each varietal and would reject samples that didn't meet true-to-type expectations.

As the world is now aware of the problems with international honey adulteration, it is more important than ever to establish baseline sensory and scientific criteria for varietal honeys. Instead, a honey on the retail shelf could be labeled as "sourwood," and yet, with no flavor profiles in place, no aroma references established by GC, and only a tiny NMR database, any honey could bear the name "sourwood." Sadly, as described well in *Vice*'s March 2020 article on honey research, some of that "sourwood" may be honey from Vietnam (Love, 2020). According to researcher Jim Gawenis of Sweetwater Labs in Missouri, who is quoted in the article, "Quite honestly, right now, Americans don't know what honey is supposed to taste like. The nice thing is, especially with the younger generations coming up, there is a lot more emphasis on looking at 'what am I eating, and why am I eating it'" (Love, 2020). Michael Roberts, executive director of the Resnick Center for Food Law and Policy, University of California Los Angeles, explains,

> I have gone to the FDA's offices multiple times for food fraud issues. I can tell you based on my own experience that the first question you're asked is: is there a food safety problem? If there is a food safety problem attached to fraud, then they're going to move on it, and they're going to move on it quickly, and they're going to be doing great things. If there isn't, then it's just not a priority. They won't say to you, 'We're not going to do anything about it,' but nothing will get done. And it's frustrating. (Love, 2020)

This is serious for the consumer who can't be sure that the honey in the bottle is the honey it says it is on the bottle. It compromises every jar of honey packed. A small company like Z Specialty relies on its reputation as a quality honey packer. Mead makers throughout the country rely on Z Specialty for pure varietals. Our dedicated customer base, some of whom have been purchasing our honey since we have been in business, is another testament to our passion.

For selecting varietal honey and for accurate labeling, we begin with the beekeeper. For example, a beekeeper moves their colonies into the orange groves of southern California about a week or so before optimum bloom in an effort to collect pure orange blossom honey. The bees are ready and waiting for the blossoms to open. When they do, scout bees locate a nectar source and fly home to share the information with their sisters. The bees do this by way of the "waggle dance," a series of movements that send out vibrations that communicate the location of abundant nectar. Over the next month or so, bees will repeatedly fly into the surrounding orchard, collect the nectar, and transform it into honey. As soon as the forage begins to fade, the astute beekeeper will close down the hives and remove the honey supers. This ensures that the nectar collected is pure orange or citrus blossom. If the beekeeper doesn't close the hive at the end of the bloom, the bees will go on to collect nectar from any forage in the area—often an assortment of spring wildflowers. The resultant honey will also be labeled "orange blossom," though it might not be quite as intense in flavor and aroma.

The artisan packer will get an assortment of samples from beekeepers and will select for the orange blossom honey that shows the strongest sensory characteristics of that variety. The next step is to ensure that these qualities are maintained throughout the packing process. Using orange blossom as the example, the characteristics to be maintained include a pale yellow color referred to as "extra white" on the Pfund Color Scale (Table 6.1) and a floral aroma and flavor with hints of jasmine, orange blossom, citrus, and even orange peel.

Artisanal packers prefer to keep the temperature of the honey low—about 105°F or 40°C. This low heat preserves the enzymes and pollen in the honey as well as the color, aroma, and flavor.

All of these can degrade if the honey is more heavily processed. A small packer uses a mesh screen of about 200 μm to remove the debris of wax, propolis, and bee parts but allows the pollen to flow through. The honey is decanted and packed into jars, sometimes with automated equipment but often by hand. At Z Specialty, we pump honey into tanks and use a semi-automatic filler. Jars are hand-fed under the filler and removed, and lids are put on by hand. After this, each jar is hand-labeled.

Z Specialty label design is another area which sets us apart from commercial packers. In the early 1980s we understood that a unique label was needed to help us sell our artisanal product. Eventually we settled on watercolors combined with pen and ink drawings, each depicting from which flowers the honey's nectar was collected. In addition, a short, colorful story describing what the fields looked like, what the honey tastes like, and where the honey came from appears on the label. Though there was a pattern to the look, each variety had a unique watercolor.

Now, with the advent of the internet, Z Specialty has put our story online. We still have our watercolor labels, though they have been reconfigured to be a bit more "modern" (Figure 6.6). An entire new line was added to the Gourmet Honey Collection called the "Golden Reserve"—a line of rare or naturally limited production honeys from across the United States.

We are still small, not only in the number of people who work for us but in the way we handle our product and think about each and every flavor and jar of honey. We now have over 30 varietal honeys from across North America. We are growing physically, too. A new 20,000–square foot facility houses our production facility, a warehouse, and a new tasting room that is open to the public.

Figure 6.6 Jars of honey offered by Z Specialty. (Left) California yellow star thistle. (Right) Golden Reserve California coriander.

Our vision for the future includes visitors stopping by to taste nature's varietal honeys, wander through a pollinator-friendly garden, watch the bees and butterflies, and picnic in our courtyard listening to live music and enjoying warm evenings.

The future offers a very sweet opportunity for a very small company.

References

Amerine, M. A., R. M. Pangborn, and E. B. Roessler. *Principles of Sensory Evaluation.* New York: Academic Press, 1965.

Biluca, F. C., F. Braghini, L. V. Gonzaga, A. C. O. Costa, and R. Fett. "Physicochemical Profiles, Minerals and Bioactive Compounds of Stingless Bee Honey (Meliponinae)." *Journal of Food Composition and Analysis* 50 (2016): 61–69.

Bogdanov, S., T. Jurendic, R. Sieber, and P. Gallmann. "Honey for Nutrition and Health: A Review." *Journal of the American College of Nutrition* 27 (2008): 677–689.

Cajka, T., J. Hajslova, F. Pudil, and K. Riddellova. "Traceability of Honey Origin Based on Volatiles Pattern Processing by Artificial Neural Networks." *Journal of Chromatography A* 1216 (2009): 1458–1462.

Castro-Vázquez, L., M. C. Díaz-Maroto, C. de Torres, and M. S. Pérez-Coello. "Effect of Geographical Origin on the Chemical and Sensory Characteristics of Chestnut Honeys." *Food Research International* 43 (2010): 2335–2340.

Codex Alimentarius Commission. *Codex Alimentarius Commission Standards.* Rome: Food and Agriculture Organization of the United Nations; Geneva: World Health Organization, 2001.

Crailsheim, K. "The Flow of Jelly Within a Honeybee Colony." *Journal of Comparative Physiology B* 162 (1992): 681–689.

Crane, E. "The Past and Present Importance of Bee Products to Man." In *Bee Products: Properties, Applications, and Apitherapy,* edited by A. Mizrahi and Y. Lensky, 1–14. New York: Springer, 1997.

Crane, E. *The World History of Beekeeping and Honey Hunting.* New York: Routledge, 1999.

Da Silva, P. M., C. Gauche, L. V. Gonzaga, A. C. O. Costa, and R. Fett. "Honey: Chemical Composition, Stability and Authenticity." *Food Chemistry* 196 (2016): 309–323.

Escriche, I., M. Kadar, M. Juan-Borrás, and E. Domenech. "Suitability of Antioxidant Capacity, Flavonoids and Phenolic Acids for Floral Authentication of Honey: Impact of Industrial Thermal Treatment." *Food Chemistry* 142 (2014): 135–143.

Escriche, I., M. Visquert, J. M. Carot, E. Domenech, and P. Fito. "Effect of Honey Thermal Conditions on Hydroxymethylfurfural Content Prior to Pasteurization." *Food Science and Technology International* 14, no. 5 (2008): 29–35.

Escriche, I., M. Visquert, M. Juan-Borrás, and P. Fito. "Influence of Simulated Industrial Thermal Treatments on the Volatile Fractions of Different Varieties of Honey." *Food Chemistry* 112 (2009): 329–338.

Escuredo, O., M. Míguez, M. Fernández-González, and M. Carmen Seijo. "Nutritional Value and Antioxidant Activity of Honeys Produced in a European Atlantic Area." *Food Chemistry* 138 (2013): 851–856.

Finola, M. S., M. C. Lasagno, and J. M. Marioli. "Microbiological and Chemical Characterization of Honeys from Central Argentina." *Food Chemistry* 100 (2007): 1649–1653.

Gámbaro, A., G. Ares, A. Giménez, and S. Pahor. "Preference Mapping of Color of Uruguayan Honeys." *Journal of Sensory Studies* 11 (2007): 507–519.

González, M. M., C. De Lorenzo, and R. A. Pérez. "Development of a Structured Sensory Honey Analysis: Application to Artisanal Madrid Honeys." *Food Science and Technology International* 16 (2010): 19–29.

Iglesias, M. T., P. J. Martín-Álvarez, M. C. Polo, C. de Lorenzo, and E. Pueyo. "Protein Analysis of Honeys by Fast Protein Liquid Chromatography: Application to Differentiate Floral and Honeydew Honeys." *Journal of Agricultural and Food Chemistry* 54 (2006): 8322–8327.

Kaškoniene, V., P. R. Venskutonis, and V. Čeksteryte. "Carbohydrate Composition and Electrical Conductivity of Different Origin Honeys from Lithuania." *LWT—Food Science and Technology* 43 (2010): 801–807.

Kelly, J. D., C. Petisco, and G. Downey. "Application of Fourier Transform Midinfrared Spectroscopy to the Discrimination Between Irish Artisanal Honey and Such Honey Adulterated with Various Sugar Syrups." *Journal of Agricultural and Food Chemistry* 54 (2006): 6166–6171.

Kristy, G. "Beekeeping from Antiquity Through the Middle Ages." *Annual Review of Entomology* 62 (2017): 249–264.

Love, S. "Your Fancy Honey Might Not Actually Be Honey." Vice, March 6, 2020. https://www.vice.com/en_us/article/884kq4/your-fancy-honey-might-not-actually-be-honey.

Majtan, J. "Honey: An Immunomodulator in Wound Healing." *Wound Repair and Regeneration* 22 (2014): 187–192.

Marcazzan, G. L., C. Mucignat-Caretta, C. M. Marchese, and M. L. Piana. "A Review of Methods for Honey Sensory Analysis." *Journal of Apicultural Research* 57 (2018): 75–87.

Moreira, R. F. A., C. A. B. De Maria, M. Pietroluongo, and L. C. Trugo. "Chemical Changes in the Volatile Fractions of Brazilian Honeys During Storage under Tropical Conditions." *Food Chemistry* 121 (2010): 697–704.

Piana, M. L., L. P. Oddo, A. Bentabol, E. Bruneau, S. Bogdanov, and C. Guyot Declerck. "Sensory Analysis Applied to Honey: State of the Art." *Apidologie* 35 (2004): 26–37.

Pita-Calvo, C., M. E. Guerra-Rodríguez, and M. Vázquez. "Analytical Methods Used in the Quality Control of Honey." *Journal of Agricultural and Food Chemistry* 65 (2017): 690–703.

Pyrzynska, K., and M. Biesaga. "Analysis of Phenolic Acids and Flavonoids in Honey." *Trends in Analytical Chemistry* 28 (2009): 893–902.

Ratnieks, F. L. W., and K. Shackleton. "Does the Waggle Dance Help Honey Bees to Forage at Greater Distances than Expected for Their Body Size?" *Frontiers in Ecology and Evolution* 3, no. 4 (2015): 31.

Siegmund, B., K. Urdl, A. Jurek, and E. Leitner. "'More than Honey': Investigation on Volatiles from Monovarietal Honeys Using New Analytical and Sensory Approaches." *Journal of Agricultural and Food Chemistry* 66 (2018): 2032–2042.

Souza, B., D. Roubik, O. M. Barth, and T. Heard. "Composition of Stingless Bee Honey: Setting Quality Standards." *Interciencia* 31 (2006): 867–875.

Subramanian, R., H. Umesh Hebbar, and N. K. Rastogi. "Processing of Honey: A Review." *International Journal of Food Properties* 10 (2007): 127–143.

Thrasyvoulou, A., C. Tananaki, G. Goras, E. Karazafiris, M. Dimou, V. Liolios, D. Kanelis, and S. Gounari. "Legislación de criterios y normas de miel." *Journal of Apicultural Research* 57 (2018): 88–96.

Tornuk, F., S. Karaman, I. Ozturk, O. S. Toker, B. Tastemur, O. Sagdic, M. Dogan, and A. Kayacier. "Quality Characterization of Artisanal and Retail Turkish Blossom Honeys: Determination of Physicochemical, Microbiological, Bioactive Properties and Aroma Profile." *Industrial Crops and Products* 46 (2013): 124–131.

Truchado, P., F. Ferreres, and F. A. Tomas-Barberan. "Liquid Chromatography-Tandem Mass Spectrometry Reveals the Widespread Occurrence of Flavonoid Glycosides in Honey, and Their Potential as Floral Origin Markers." *Journal of Chromatography A* 1216 (2009): 7241–7248.

US Department of Agriculture. "United States Standards for Grades of Extracted Honey." *Federal Register* (1985): 50FR15861.

Wahdan, H. A. L. "Causes of the Antimicrobial Activity of Honey." *Infection* 26 (1998): 26–31.

Wang, J., and Q. X. Li. *Chemical Composition, Characterization, and Differentiation of Honey Botanical and Geographical Origins.* Amsterdam: Elsevier, 2011.

Wang, X. H., N. Gheldof, and N. J. Engeseth. "Effect of Processing and Storage on Composition and Color of Honey." *Journal of Food Science* 69 (2004): 96–101.

White, J. W. "Honey." *Advances in Food Research* 24 (1978): 287–374.

Yaghoobi, R., A. Kazerouni, and O. Kazerouni. "Evidence for Clinical Use of Honey in Wound Healing as an Anti-bacterial, Anti-inflammatory, Anti-oxidant and Anti-viral Agent: A Review." *Jundishapur Journal of Natural Pharmaceutical Products* 8 (2013): 100–104.

7

INDUSTRIAL AND ARTISANAL OLIVE OIL

MIKE MADISON

Introduction

Although there are no formal definitions of industrial and artisanal olive oils, it is usually not difficult to assign a producer to one category or the other. Typically, the industrial producer is a corporate entity operating on a large scale and sourcing olives from many different farms, while the artisanal producer is an individual working on a small scale with olives from a single grove. The industrial producer has one product—olive oil—whereas the artisan is likely to make several oils that differ in flavor and intensity according to the cultivar of olives and their maturity at harvest. In the trade-offs between yield and quality, the industrial producer will sacrifice quality in favor of yield, while the artisan will accept decreased yield for better flavor.

It is instructive to imagine the conditions and processes—growing, harvesting, and milling—which would produce the best possible olive oil within the confines of present-day technology.

Mike Madison, *Industrial and Artisanal Olive Oil* In: *The Science and Craft of Artisanal Food*. Edited by: Michael H. Tunick and Andrew L. Waterhouse, Oxford University Press. © Oxford University Press 2023. DOI: 10.1093/oso/9780190936587.003.0008

Having done this, we can examine current industrial and artisanal practices of olive oil production, noting the issues of scale, economics, technology, and philosophy that lead to deviations from this ideal and result in the production of olive oil that is chemically and sensorially less than what it could have been.

The Best Possible Oil

In the ideal harvest, the trees in the grove have been pruned to an open configuration so that abundant light reaches the interior of the tree. Deficit irrigation leading up to harvest has raised the polyphenol content of the olives and lowered their water content. Some of the fruit is turning color, some of it is still green—all of it is in perfect condition. It is gently hand-harvested without damage and placed in shallow, well-ventilated boxes, which are immediately taken to the mill located just meters away.

At the mill, the olives are dropped past a fan that blows away leaves and then washed in clean water. The olives immediately are crushed in either a hammer mill or a disc crusher, the loading rate slow enough to prevent temperature rise. From this point forward, all operations are in an oxygen-free environment. The crushed olive paste is moved to the malaxer, where it is slowly stirred for about 30 minutes. Four things are happening during malaxation: added pectinase enzymes are breaking down cell walls to increase release of the oils; the temperature is slowly being brought up to the working temperature of 24°C; microscopic droplets of oil are aggregating into bigger and bigger oil drops, which facilitates separation; and polyphenols and flavor elements that in the cell had been separated from the oil at a cell-organelle level are now combining with the oil—that is, the oil is acquiring its flavor.

The olive paste is pumped into a two-phase horizontal decanter centrifuge with no water added. The decanter is operated at only about two-thirds of its rated capacity, increasing the dwell time in the decanter, which results in a cleaner separation of the oil from the pomace. The oil is moved to a tall, narrow tank to allow separation by gravity of any water mixed with the oil; the majority of the water settles out very rapidly. After an hour or so in the tank, the oil is filtered through a cellulosic filter that removes tiny particles of olive flesh and any remaining water. The filtered oil is transferred to a stainless steel storage tank with a headspace of inert gas (nitrogen or argon) at a comfortable temperature of about 18°C.

What is the nature of this ideal oil? To the eye it is clear and a lovely deep green color, due to chlorophyll that was in the skin of the olives. The oil has a fruity fragrance when sniffed. In the mouth it is bright, clean, and fresh with no waxy or greasy mouthfeel. The flavor in the forward part of the mouth is floral/fruity/vegetal; toward the back of the mouth one tastes distinct bitterness, which in olive oil is considered a positive trait; and in the throat the oil is pungent and peppery, eliciting a cough.

Chemically, this ideal oil is about 99% triacylglycerols. The fatty acid moiety of these triacylglycerols is primarily oleic acid but includes a number of others, including polyunsaturated and saturated elements as well as the more abundant monounsaturated fatty acids. There is quite a bit of variability in the relative abundance of these compounds (Table 7.1). For example, palmitic acid varies between 7.5% and 20% of the total, depending on the sample. These differences reflect genetic differences among cultivars,

Table 7.1 Fatty Acid Composition of Olive Oil, Range
of Percentages in Various Oils

Fatty acid	Percentage
C14:0, myristic	0.0–0.1
C16:0, palmitic	7.5–20
C16:1, palmitoleic	0.3–3.5
C17:0, margaric	0.0–0.5
C18:0, stearic	0.5–5.0
C18:1, oleic	55–83
C18:2, linoleic	3.5–21
C18:3, linolenic	0.0–1.0
C20:0, arachidic	0.0–0.8
C22:0, behenic	0.0–0.3
C24:0, lignoceric	0.0–1.0

Source: Boskou et al. (2006).

growing conditions (irrigation, temperature, light, maturity), and conditions of milling. Because the triacylglycerols are essentially flavorless, even large differences in their composition among oils affect flavor only slightly or not at all.

It is the 1% of the oil that is not triacylglycerols that determines the flavor, and many of the health benefits, of olive oil. About half of these compounds are polyphenols (tyrosol, hydroxytyrosol, oleocanthal, oleuropein, and others). These contribute to the bitterness and pungency of the oil. In addition to these there are hundreds of other chemical compounds present in the oil: sterols, squalene, flavonoids, terpenes, pigments, etc. Boskou et al. (2006) cite references to 280 different volatile compounds, mostly sampled from the headspace of bottled oils, that are found in olive oil and contribute to its fragrance and taste.

Because of the great diversity of chemical compounds in olive oil, most of which are present only in minute quantities, it is not feasible to evaluate oil by laboratory analysis alone. In addition to laboratory measurements, olive oil is formally evaluated and judged by sensory analysis (i.e., tasting). While olive oil judges are trained to adhere to specific standards, their conclusions nonetheless reflect a large element of subjectivity.

Industrial Olive Oil

Producers of industrial olive oil operate under three constraints: they compete based on price, they operate at a large scale, and they can never empty their tanks and run out of oil because the demand for their products is continuous. These constraints cause the producers to deviate from the ideal production methods in several ways.

Traditionally, the most expensive part of olive oil production was hand-harvesting the fruit. In the last few decades we have developed mechanical harvesters of two main types. In one type, the trees are planted at very high density (spacing of 1.5 × 3 m) and pruned to a hedge. The harvesting machine straddles the hedge and knocks off the olives with rotating fiberglass rods. The fruit is picked up on a conveyer and shunted to a bin. This is a violent process and works best when the fruit is quite green, so harvest may be as early as September. Even then, much of the fruit is bruised in the process. Only a few cultivars, primarily Arbequina but also Arbosana and Koroneiki, are suited to this method. The other mechanical harvesting technique uses equipment developed for harvesting prunes and pistachios. A hydraulic arm grasps the trunk and shakes the tree powerfully at a variety of frequencies, and the olives that are shaken from the tree fall onto catch frames and are moved to a bin. The shaker damages the bark on young trees and is suitable only for trees of about 10 years old or more. A full array of harvesting equipment is expensive, well over $500,000, which puts it out of reach of artisanal producers. However, operated on a large acreage, the per-acre cost is not great, which allows the industrial producers to harvest at a much lower cost than hand-harvesting.

It is worth noting that the olives on a tree do not ripen simultaneously. Ripening begins at the very top of the tree, then on the southeast side of the outer branches (in the northern hemisphere), and much later in the deep interior of the tree. Heavy pruning to keep the interior of the tree open makes ripening somewhat more uniform. As an olive ripens it begins to form an abscission layer at the point of attachment so that eventually it will fall from the tree. Harvesting with a mechanical shaker favors the ripest fruit, which is most easily removed, and leaves greener fruit on the tree. The shaker also preferentially harvests damaged fruit. This is because damage to the fruit (fly larvae, fungal infection, bird pecks, branch abrasion) triggers a wound response that accelerates maturation, including formation of an abscission layer. The upshot is that from any given tree an artisanal hand-harvester will select better fruit than what is harvested by the shaker. Early in the season when the olives are green, the shaker may harvest only 60%–70% of the olives. Because of this, mechanical harvest with a shaker is often delayed to a later stage of maturity when the percentage of fruit harvested is higher, with the result that the oil produced is likely to be milder and less robust than it would have been from an earlier harvest.

Because the industrial producer works on a large scale, some of the olives processed will be coming from a distance, so the olives must be transported. When olives are placed in a bin (60 cm deep) or in a dump vehicle (2 m deep), the fruit on the bottom is crushed and begins to ferment. This effect is amplified to the extent that the fruit must be transported down a bumpy road. When olives are transported 100 km on a warm day, the fruit on the bottom of a bin may already be at a temperature of 40–45°C by the time it gets to the mill, due to fermentation and oxidation. Despite careful scheduling, it often happens that deliveries are backed up at the mill yard; and a delay of several hours, or even days, before milling can contribute to degradation of the oil.

In the mill, the operator may follow practices that increase yield at the expense of quality. This includes increasing malaxation time, malaxing at high temperatures (30 or even 34°C) rather than the ideal temperature of around 24–25°C, and adding water to the olive paste as it is pumped into the decanter. Higher malaxation temperatures alter the oil chemistry and notably diminish its sensory quality (Angerosa et al., 2001). When the oil emerges from the decanter it may be cleaned in a separator (vertical centrifuge). In this process warm water is used to wash the oil and clarify it by removing particles. Again, this involves a loss of polyphenols; the advantage is that the oil is more easily handled and stored. The separator is faster and more economical than filtering with a cellulose filter.

The industrial producer cannot run out of oil without serious consequences. The producer is at the head of a long supply chain, and if the contracts cannot be fulfilled, the buyers will find another source. Once broken, these relationships are difficult to re-establish. So the industrial producer always has oil in storage, often for more than 1 year. Even under the best conditions olive oil slowly degrades in storage; ideally, it should be used in the year of production. Industrial olive oil may be old and partially oxidized by the time it is sold.

Industrial producers often purchase oil to maintain their stocks, and one option is to purchase defective oil inexpensively with the intention of rectifying the defects. The defects may arise before the olives are milled due to fungal or bacterial infection, prolonged storage, olive fly damage, or excess heat. Damage may occur during milling due to high temperatures. And defects may result from storage of the oil due to exposure to heat, light, and oxygen or from prolonged contact with water and fruit particles that were not separated from the oil and harbor bacteria, either aerobic or anaerobic.

The commonest defect is classic rancidity that results from oxidation of the triacylglycerols, yielding free fatty acids. But in addition to rancidity, oxidation may act on other chemicals present in the oil to produce off-flavors. If olives are heavily infested with the olive fruit fly, fly larvae in the fruit may add a disagreeable flavor to the oil. The oil can be cleansed of these defects by some combination of steam distillation, treatment with charcoal, and extraction with solvents, resulting in refined oil that no longer has the flavor defects but has also lost all of the beneficial polyphenols and other flavor elements of good oil. The final product is a colorless, tasteless, odorless, oily fluid. The oil profile has been modified as well, with an increase in diacylglycerols and a shift in the proportions of the fatty acid moieties.

Unscrupulous producers may blend refined olive oil with virgin oil, possibly adding chlorophyll as a coloring agent. In addition to dilution with refined oil, olive may be adulterated with oils from other plant sources such as canola or hazelnut. The vast extent of fraud in olive oil of Mediterranean origin became widely known with the publication in 2011 of the book *Extra Virginity* by Tom Mueller. The majority of US supermarket oils of Mediterranean origin are mislabeled "extra virgin" and are rancid. It is not illegal to refine olive oil, to blend it with oils from other sources, or to sell rancid oil; what is illegal is to label these oils as "extra virgin olive oil." We have chemical tests that can detect and measure various defects including rancidity, adulteration, and blending with refined oils so that fraud can be easily discovered. Nonetheless, enforcement of labeling laws continues to be lax.

A curious result of this is that many Americans believe that rancid oil is the correct taste of olive oil, and if they are offered a good-quality extra virgin oil, they perceive it as defective because it fails to conform to their prior experience. This has given the producers, both the artisanal ones and the honest industrial ones, the Herculean task of re-educating the public's palate through tastings, lectures, and demonstrations.

Most of the industrial producers have just a single product: olive oil. In a few cases, a monocultivar oil will mention the cultivar name on the label (usually Arbequina or Koroneiki), but generally there is no mention of cultivar names because the oil is made from an unquantified blend of olives from many sources. One of the skills required of the industrial producer is to blend oils from many different sources in a way that maintains a consistent product. This is unfortunate in that it fails to express the great variations in flavor possible with olive oil. It suggests that olive oil is simply a fungible industrial commodity differing only in name from corn oil or soybean oil. It is as if wine were just labeled "wine," with no mention of the type of grape from which it was made.

Artisanal Olive Oil Production

The most expensive part of olive oil production is harvesting the olives, and the small-scale producer cannot easily take advantage of mechanical harvesting technology. The high cost of a suite of machines cannot be justified for a small acreage. The grower might instead hire a harvesting contractor; because the olive harvest is later in the season than prune and pistachio harvest, tree-shaking equipment is available for hire. But this raises another problem: the artisanal grower most likely has a smaller mill capable of handling only a few tons of olives per day. In contrast, the mechanical harvester has an output of 15–20 tons per day or more. And so the grower must find a contractor willing to bring in a crew for only an hour or two each day and then leave the equipment in the grove until the next day. Alternatively, the artisan must purchase a needlessly large and expensive mill capable of handling the output of the mechanical harvester. There are some less expensive shaker-harvesters available that unfold an inverted umbrella around the base of the tree to catch the fruit. This mechanism has a reputation for being temperamental and unreliable and requires a much wider spacing of the trees than is usual in most modern plantings.

Because of these issues, most artisanal producers hand-harvest their olives. While slow and expensive, hand-harvesting also permits careful selection of the fruit. Indeed, this is the artisanal producer's primary advantage: complete control of the growing and harvesting of the olives from which to make the oil. The industrial producer sourcing olives from many providers has very little control over the quality of the olives, and if quality of the fruit is mediocre, then mediocre oil is the best that can be made.

Pruning and irrigation of the trees form the foundation for excellent olive oil. Pruning achieves a number of goals: it facilitates harvest by keeping fruiting branches accessible; it makes a healthier tree by letting light and air into the interior, which discourages both the olive fruit fly and the scale insects that supply the adult flies' food; it promotes a more uniform ripening of the olives by getting light to the fruit in the interior of the tree; it promotes drought tolerance by changing the ratio of shoots to roots in favor of roots; and it helps to even out alternate bearing, in which the trees bear a heavy crop every second year and a scanty crop in the off years.

While the trees should be well irrigated at flowering time in order to assure good fruit set, deficit irrigation is appropriate for the remainder of the season. Deficit irrigation elicits a stress response from the trees that includes increased synthesis of polyphenols and other secondary compounds which wind up in the oil as important elements of flavor. Overirrigated trees are likely to produce bland oil. On the other hand, excessive drought stress may result in excessively bitter oil. The cultivar Coratina is notorious for extremely bitter oil when insufficiently irrigated.

The artisanal producer who does not own a mill must take the olives to a public mill that provides fee-for-service milling. Scheduling may be difficult, especially in years of a heavy crop, so that the artisan may not be able to harvest on a preferred date. Most public mills have a minimum weight, for example, one ton or five tons; and this may force the artisan to blend olives that they would have preferred to keep separate in order to meet the minimum weight requirement. And the artisan must depend on the miller to comply with requests for malaxation temperature and time, decanter dwell time, addition of water, and so forth.

For these reasons, it is preferable to have one's own mill. Although there are very small mills available, their design involves considerable compromise; the smallest mill that offers the full array of control options would be one with a capacity of about one-half ton per hour. Such a mill will cost $100,000 or more, with another $100,000 for a building to put it in, not to mention all the permits and licenses that go along with that. The artisan who is short of capital may do fee-for-service milling in order to pay for the mill; in fact, custom milling is a far more profitable business than making and selling olive oil.

The artisanal producer who owns a mill is able to adhere closely to the ideal regimen described at the beginning of the chapter. The artisan is also in a position to experiment freely with the many variables of harvesting and milling; that knowledge, slowly acquired, allows excellent oil to be made in a predictable way.

Artisanal Olive Oil Marketing

The industrial producers enjoy genuine economies of scale. At every step of the process of oil making, the artisanal producer has greater capital costs and greater operating costs per unit of oil. Even with such a simple transaction as purchasing bottles, the industrial producer who buys 100,000 bottles at a time will pay a notably lower price than the artisan who buys only 1,000 bottles. A consequence of this is that the artisan cannot compete on price, so they must compete on quality.

The first requirement for marketing based on quality would seem to be to actually have good-quality oil: complex, interesting, and free of defects. Olive oil judges look for four attributes: fruitiness, bitterness, pungency, and a harmonious balance among these traits. Compared to industrial olive oils, nearly all artisanal olive oils will be judged to be of a much higher quality.

Judging oil is not straightforward. One wishing to become an olive oil judge is trained by other judges to evaluate oil in a particular way according to criteria which are at least somewhat arbitrary. So, in a sense, becoming a judge amounts to indoctrination into a culture. The criteria of judging differ slightly from region to region, and no two panels of judges are likely to reach the same conclusions when judging an array of oils. Where the panels are in agreement is in their ability to detect defective oils. In contrast to the judging of wine, cheese, coffee, and chocolate, judging oil entails a highly artificial experience—drinking the oil. Hardly anyone other than judges drinks shots of olive oil; the oil is invariably mixed with food. And so an oil that is down-scored by a judge for being too harsh might, in fact, be perfect when added to a rustic dish of beans, pasta, and bitter greens.

The artisanal producer has a number of ways to signal that their product is of high quality, thus justifying its high price. Packaging—dark green glass bottles with excellent labeling and graphics—can indicate higher quality. Some producers overdo this to the point of using extreme packaging—bottles so tall and narrow that they won't fit on an average shelf; like the unusual bottles of very expensive perfumes, the bottles themselves are meant to suggest precious contents. While industrial olive oil typically is sold in units of 1 liter or more, most artisanal oil is sold in bottles of 500 mL, or even 375 mL. The cost per liter of the artisanal oil is much higher, but the cost of a single purchase is similar to industrial oil because the artisanal oil is in a smaller container.

The use of the term "extra virgin" is supposed to be an indicator of high quality. The term has a long and nebulous history and now has legal definitions, different in different countries, that include both laboratory measurements (e.g., free fatty acids <0.8%) as well as standards of sensory evaluation. The chronic fraudulent use of the term "extra virgin" on low-grade oils has greatly diminished its utility, and it is further devalued by seed oil producers co-opting the term (e.g., "extra virgin safflower oil").

Certification is another method of suggesting higher quality. Both the California Olive Oil Council and the Australian Olive Association offer supplemental labels that certify quality. Certification of organic growing (one certificate) and organic processing (a second certificate) suggest careful production, and the organic label is attractive to many buyers. There are other certifications for fair trade, sustainable harvest, neutral carbon footprint, and so forth. In some European regions there are certifications of geographic origin (protected designation of origin and protected geography indication) which may be indicative of oils of high quality. There have been several attempts to establish independent agencies (Tre-e, Extra Virgin Alliance, Ultra-Premium) that certify oils based on their adherence to chemical and sensory standards more stringent than the legal definitions of extra virgin, but these have not been widely adopted.

There are many olive oil competitions where one may submit oils for judging. Bronze, silver, and gold medals are given, as well as designations "best of class" and "best of show." It is not difficult to win a gold medal—they are handed out quite freely with no limit to the number of them, and many producers collect these medals as symbols of high quality to be used in advertising and on their labels. So far olive oil has been spared the insidious practice of scoring oils on a scale of 1 to 100, though it is likely to show up sooner or later.

Many people use price as a surrogate of value so that a high price may be used to signal exceptional quality. Scarcity may act similarly. If it is advertised that there are only 48 cases of this oil available and when those are gone that's the end of it, some people will automatically assume that, because the oil is rare, it is special and of high quality. Scarcity is not necessarily a gimmick. Some oils (Ascolano, Sevillano) are so low in polyphenols that they have a short shelf life; they should be offered for only the first few months after harvest.

The above are standard marketing ploys about which one might well be cynical, but there is another practice with implications for marketing that is valid and important: offering a variety of olive oils based on the olive cultivars from which they are made. Consider for a moment the situation with wine. Wine has four variables: brand, appellation, vintage, and grape cultivar. The interaction of these creates complexity; complexity creates interest. Olive oils will not have vintage since they are used in the year of production, although indicating the harvest date provides a useful clue to the intensity of the oil. So far, the development of appellation in California is still immature; we do not yet know, for example, if a Frantoio oil from the central coast differs in a consistent way from a Frantoio oil from the Sacramento Valley. So the boundaries of appellations are still emerging, though it is likely that appellation will eventually come to be an important aspect of artisanal oils. The most important variable, which is underutilized, is the olive cultivar from which the oil is made. A delicate Ascolano oil is as different from a robust Coratina oil as a Chardonnay wine is from a Malbec. To fail to indicate the cultivar names on the label is to deprive the buyer of important information. We are still at the beginning of a long process of educating consumers who previously considered olive oil to be a single fungible product, and a rancid one at that. Using varietal names alerts the users to the idea that varietal oils differ, and this in turn stimulates them to pay closer attention to the taste of the oil.

Artisan Philosophy

At this point I'm switching to a first-person narrative, which seems appropriate for describing my own artisanal operation. I began planting olive trees in 1991 at a time when artisanal olive oil was just getting started in California. I was acquainted with many of the other growers; we attended the same workshops and short courses and seminars. It struck me that most of them were older, wealthy people who had successful careers in the professions or business and were now looking for something useful and interesting to do. Making olive oil requires skill, knowledge, and judgment and carries considerable cachet. It is on a par with winemaking. I suspect that some of them wanted to recreate the landscapes and leisurely ambiance of fondly remembered holidays in Italy or Greece. And many were frustrated by the unavailability of good olive oil in California. All of us were determined to make excellent oil.

At one of those early gatherings, a prominent producer remarked, "I hope that someday we will be competitors, but for the time being we must all be colleagues." What he referred to was the shared task of educating an uninformed populace about what good olive oil is, as a prerequisite to developing enough demand for us to become competitors.

I don't think that most of the early artisans were motivated primarily by a desire to make money. If that were the case, a cursory glance at the data would show that almost any other tree crop would be a better choice than olives. Nonetheless, pretty much all of them embraced the 20th-century corporatist philosophy of maximizing efficiency, charging the highest prices that the markets would tolerate, and maximizing profits. They used expensive and tasteful packaging, and as soon as competitions became common, they were assiduous about collecting gold medals and the bragging rights that go with them. (This is to some extent surmise, and if I have misjudged anyone, I apologize.)

My situation was different. Our income came from operating a small (9 ha) farm, which kept us above poverty level but not by much. The appeal of olives was the timing of harvest—November and December. This was a time when our other crops were dormant, and we had little work to do and no income. Olive trees are easy to grow: they are drought-tolerant and pest-free (at that time the olive fly had not yet appeared in California), and they do not require bees for pollination. In addition, I had an aesthetic motive: olive trees are strikingly handsome plants, and the flash and sparkle of their silver foliage on a windy day was reason enough to plant them.

At first we more or less followed the corporatist business ideas of the other growers; but that soon changed, and over the years we have developed our own philosophy, or at least so it seems in retrospect. This began with the question, "Who buys a half-liter bottle of oil that costs $40?" The answer is "rich people" (or maybe "gullible rich people"). To serve only the wealthy was at odds with our values and our social identity. And so for a long time we struggled with the issue of pricing. Most of our sales are at the weekly farmers' market in a nearby town of about 100,000 inhabitants, and while some of our customers are wealthy, many are struggling working people or students on tight budgets. We tried to come up with a pricing schedule that was fair to our customers and to ourselves but without success. In an unjust society, there is no just price. And so we decided to set our prices at the very bottom end of the price scale of artisanal oils, a price which to some extent overlaps the high end of industrial oil prices. And we changed our bottling so that while we still offer half-liter bottles, most of our oil is sold in 3-liter jugs, which greatly decreases the proportional cost of packaging. We have trained our customers to use the oil freely so that most of them will use up a jug in 4 or 5 weeks. A problem we had observed with small, expensive bottles of oil is that people consider the oil to be "too good to use," so they hoard it, whereas with a large jug, the cook feels rich in olive oil and uses it generously. In addition to selling oil, we barter it for wine, cheese, avocados, and veterinary care.

We do not advertise or enter competitions, nor do we seek out any certifications other than organic certification. Those are useful to producers who are attempting to sell oil to strangers, especially producers who specialize in sales on the internet. But we know nearly all of our customers by face, if not by name, and they know us. And so there is a personal relationship in which formalized bragging is unnecessary.

On the farm I believe that care of the trees is the most important task. The very best that a skilled miller can do is to refrain from making the oil worse. Poor oil cannot be made better. So I strive to produce the best fruit possible. I prune the trees very hard, removing about 20% of the wood each year. These trees are growing on deep, fertile soil and are very vigorous. In tougher soil, less pruning would be required. I am very frugal with irrigation, with the result that the polyphenol content of the oil is high, and the oils are very robust. At the mill I favor quality over yield. I mill at a low temperature (24°C) and at a rate well below the maximum capacity of the decanter. I plan to add a filtering system in the coming year.

We have no employees on the farm. The regulatory burdens and costs imposed by the state of California make that impossible; it is also true that I enjoy the solitude of work or, even better, the companionship of crows and rabbits. I hand-harvest the fruit by myself, starting in the second week of October and aiming to finish by the end of the year. Usually I harvest during the day, placing the olives in shallow, well-ventilated crates and bringing them to my mill, just 60 m distant from the grove. I mill the olives the next morning at 4:30 a.m., finishing before sunrise, which gives me time to drink a cup of coffee and read the newspaper before the start of the day's harvest. We have about 1,700 trees representing 16 cultivars. Five hundred trees are leased out to some young friends, and some we don't harvest; so I end up harvesting somewhat more than 1,000 trees and producing a few thousand liters of oil. By the first of November when we have new oil in hand, we donate any oil remaining from the previous year to agencies that feed indigent people, thus freeing up our tanks for incoming oil.

Like many artisans, our philosophy and methods of operation are essentially medieval; they would not be out of place in 15th-century Spain. With a high input of manual labor we produce the best oil that we can and sell it nearby, at a modest price, to people we know. In a region that has many such artisans (winemakers, cheesemakers, coffee roasters, chocolatiers, market gardeners, butchers, bakers, chefs), the result is a healthy and vital community with strong social cohesion, altogether superior to those impoverished places where commerce is limited to the trade of anonymous industrial products.

References

Angerosa, F., R. Mostallino, C. Basti, and R. Vito. "Influence of Malaxation Temperature and Time on the Quality of Virgin Olive Oils." *Food Chemistry* 72 (2001): 19–28.

Boskou, D., G. Blekas, and M. Tsimidou. *Olive Oil*. Amsterdam: Elsevier, 2006.

Mueller, T. *Extra Virginity*. New York: W. W. Norton, 2011.

8

ARTISANAL FRUITS AND VEGETABLES

ROSEMARY E. TROUT AND JOSEPH J. TROUT

Rationale for Artisanal Production

Production of agriculture is ever-changing but continues to be a large segment of a national economy. The way food is produced, packaged, and marketed is also changing. Consumers are increasingly interested in how their food is produced and where it comes from. There is a desire to purchase and consume produce that is ethically and sustainably farmed. *Artisanal* is a widely used term regarding varieties of foods, including breads, chocolate, cheese, ice cream, coffee, vanilla, bread, and meats (Cope, 2014; Hartman, 2015). It is often used in conjunction with or in the same vein as terms like *heritage, heirloom, ecological, handcrafted, handmade,* and *specialty*. Some studies show that consumers prefer artisanal food and are willing to pay higher prices for it (Cope, 2014). It is associated with better flavor, nutrient density, and support for small-scale, family-owned farms that use traditional, non-commercialized

Rosemary E. Trout and Joseph J. Trout, *Artisanal Fruits and Vegetables* In: *The Science and Craft of Artisanal Food*. Edited by: Michael H. Tunick and Andrew L. Waterhouse, Oxford University Press.
© Oxford University Press 2023. DOI: 10.1093/oso/9780190936587.003.0009

methods of farming. In addition, the psychological aspects associated with the preference, or acceptance, of artisanal foods are partly based on their chemistry (Cirne et al., 2019).

In the medieval period and through the Renaissance, the term *artisan* was applied to one who studied with a known master who was the artist. The artisan practiced and honed their craft under the direction of the artist. Throughout history, prior to industrialization, the farmer could be considered to be an artisan since young farmers studied under experienced farmers to hone and perfect their farming skills. Post-industrialization, the meaning of *farmer* has changed; now, it can be applied to anyone who is involved in the growing process, whether it be large-scale conventional commercial farms, small rural farmers, or urban community garden growers.

The term *artisan* has evolved and taken on the meaning of one who makes something by hand, with specific knowledge of the process, and who creates, grows, or produces their product with traditional methods. The term *artisan* can refer to *who* is producing something, and *artisanal* can be applied to *what* they are producing and *how* it is produced. Traditionally, *artisan* refers to both *how* something is made and of *what* it is made. Dictionaries define the term as someone who does skilled work with their hands. In common usage in food, it more often refers to how something is made or produced, thereby making it somewhat of a marketing term. In food, *artisanal* as a term is used freely on packages and as a descriptive term, with no associated federal standard identity associated with it (Cope, 2014; Mapes, 2020). Highly mechanized farms may use the term as readily as a small farmer who tills and sows their field almost completely by hand or a local grower in an urban community garden who sells their produce at the local farmers' market. In essence, it has become a marketing term with no consistent meaning but with a connotation that infers that the food product is made or traditionally farmed by hand, with smaller yield and minimal processing, even though this may not actually be the case. The term may also increase the price of the product in various stages along the supply chain (Cope, 2014). A federal standard of identity or standards of common usage of the term *artisan* among stakeholders in a particular segment of the food industry would be beneficial for various reasons: consistency in meaning for producers and marketers and better understanding for the consumer.

Differences in Production Methods

Conventional, organic, and sustainable artisan agriculture are three categories with which to organize cultural farming practices in the United States (Asami et al., 2003). Each has its own goals and methods, with some overlap between the three. Each also had its advantages and disadvantages. Conventional farming focuses on high yield and lower prices in the retail marketplace; however, consumers are looking for more sustainability and reduced negative perception of land use, carbon emissions, and environmental impact. Organic farms must specifically employ standards and practices set forth by the US Department of Agriculture's National Organic Standards (Bourn and Prescott, 2002). These are strict standards that must be met without deviation and require land used for organic farming to be free from the use of any synthetic pesticides or herbicides for 3 years prior to harvest. Crop management is often done through rotation, planting disease-resistant cultivars, and using animal manure as fertilizer. No genetically modified seeds or plants may be used, nor can irradiation or fertilization from sewage. Sustainable or artisanal farming is focused on traditional methods, heirloom cultivars, and heavy consideration for consumer preference for environmentally healthy, ethical, and equitable farming in a specific region (Asami et al., 2003).

Fruits and vegetables are available in many forms in the retail market: fresh whole, fresh cut and sold in modified atmospheric packaging, dried, frozen, and canned. There are many differences between large-scale commercial fruit and vegetable farming and small-scale sustainable, artisan farming, from pre-production through post-production, shipping, and retail sale (Cirne et al., 2019). As defined by the US Department of Agriculture, conventional farming methods can include "the use of seeds that have been genetically altered using a variety of traditional breeding methods, excluding biotechnology, and are not certified as organic." Conventional crops may be grown and sold as commodities, or they may be mixed in with other crops that may be genetically modified. This is not a new practice, and conventional plant breeding has been practiced for thousands of years (US Department of Agriculture, 2015). Organic standards notwithstanding, large-scale commercial farms generally allow the use of chemical pesticides, herbicides, growth hormones and/or regulators, seeds that may have been genetically modified, and large-scale equipment in all stages of the growing cycle through packaging and shipping. Precision-style farming is a method of large-scale commercial farming. The precision-style method of farming is based on the concept of dividing larger fields into smaller parcels of land that can be managed by observing, sampling, measuring, and responding to variables specific to each parcel of land. When aggregated, the entire farm is managed using accumulated precision crop methods. Specific targeted management of particular pests and diseases is also common, through integrated pest and disease management (IPDM) (Zhang, 2018). An example of IPDM is infestation and pest modeling based on weather data, seasons, insect cycles, or other variables (Zhang, 2018). Good agricultural practices (GAP) are set for both conventional and artisanal farms. An example of GAP includes proper hygiene and sanitation for farmworkers (Sinha, 2012).

In smaller-scale, artisan farms, including farms growing heirloom crops, crops grow over smaller parcels of land. Depending on the culture of the farm, chemical pesticides, herbicides, and growth hormones and/or regulators may or may not be used. In the management of smaller farms, there is a stronger focus on seasonality. These artisans plant with the seasons, likely rotate crops, and simultaneously grow more variety of produce within a single farm, using traditional, pre-industrial farming methods. Handpicking and hand sorting are standard, and produce is harvested when ripe. Growing seasons are shorter and hyperseasonal. Artisan produce may be sold at farmers' markets, as part of a farm-share directly to consumers, in retail food markets, or to restaurants. Flavor, seasonality, sustainability, and a desire to preserve traditional methods are the artisanal farmer's main priorities. Hydroponic and biodynamic methods used in these farms are along the same lines as the artisan farms; however, some hydroponic and biodynamic farms, especially the hydroponic farms, can become quite large (Macieira et al., 2021; Sánchez et al., 2021; Heimler et al., 2012).

Harvest

Smaller-scale artisan-style farms rely on handpicking and lower yield based on less acreage and fewer precision crop management techniques, due to cost, and focus on other priorities, such as seasonality and sustainability, as contrasted with yield in conventional farming. Smaller-scale artisan farmers rely more on traditional farming techniques, often organic methods, and focus on sustainability, crop rotation, or biodynamic farming. Biodynamic farming includes blending of different varieties of annual and perennial fruits and vegetables, other plant species, insects such as bees and earthworms, and sometimes animals on the same farm. One of the main goals of the biodynamic farmer is to increase diversity and improve the soil in this traditional method (Biodynamic Association, 2021).

Large-scale farms in California that grow stone fruits like plums, peaches, nectarines, and table grapes do so on tens to hundreds of acres, planting each tree or vine as close to the next as possible while maintaining maximum sunlight exposure. To accomplish this, trees may be tied or wired open so that branches are spread apart to get sunlight into the interior stems and leaves, which will maximize yield (Zhang, 2018). Single types of crops are planted together, and there tends to be less variety within each farm. Growing seasons, especially in warm, sunny states like California, Georgia, Arizona, and to some extent Florida, are longer than each individual season. If the weather is favorable, planting and harvest cycles continue throughout the year, instead of simply during the traditional harvest season for a specific type of produce. Strawberries, once only grown and harvested in spring, are available all year in the United States (Sinha, 2012). So are apples, although the apples picked in fall are from the harvest the same year. Apples in the market in spring and summer were likely harvested during the prior fall season. They are held under tightly controlled atmospheric conditions so that they are saleable throughout the year (Zhang, 2018; Sinha, 2012).

Large-scale commercial farms plant one to a few types of agricultural products in an area. The seeds are carefully chosen, tested, and bred for specific characteristics desirable in the produce. Examples are color, size, hardiness during transit, consistency from one growing season to the next, or resistance to pests. Farming efficiency is primary, and yield is high to meet the demands of consumers who want fresh produce through the year without considering seasonality (Cope, 2014). These agricultural products are available in most of the United States, at a lower unit price, throughout the year. Much of this produce is sold through food retail chains via wholesalers as fresh, frozen, canned, and to a far lesser extent dried produce. Some is shipped to other countries, and some is further processed as ingredients in formulated foods, ultimately sold to the consumer through retail food chains or used in industrial settings such as schools and hospitals.

After produce is harvested, usually during a very brief and specific time frame, it is washed and cooled, sorted, graded, packaged, and shipped. All of this happens within hours of harvest (Zhang, 2018; Sinha, 2012). Most commercial fruits and vegetables are harvested while still immature. Peaches, plums, nectarines, apples, and avocados are examples. These fruits will travel many miles in shipping or by train and will ripen off the vine while in transit. Some of this produce may be vine-ripened, which means harvested when naturally ripe and still attached to the tree, vine, or bush. The price per unit for vine-ripened fruit is higher, but there is an appeal for these types of products because they seem less processed and more natural.

Bananas are a special case in that they are shipped at cold temperatures to the United States while still green and at less than 75% maturity (Robinson and Saúco, 2011). Upon arrival at wholesalers, they are held in warehouses where they are artificially ripened with ethylene gas prior to shipping to a final destination for sale or use. Bananas grown locally in tropical climates are typically harvested at 90% maturity and sold locally (Robinson and Saúco, 2011). While access to many varieties of most fruits and some vegetables is increasing, there is only one variety of banana, the Cavendish, commonly available for retail sale in the United States (Robinson and Saúco, 2011). Contrast the single variety of bananas with the large variety of apples. You can easily find a plethora of apple varieties, including McIntosh, Granny Smith, Pink Lady, Fuji, Honeycrisp, Gala, Red Tango, Jazz, Pacific Rose, and Red Delicious (Pollock and Crassweller, 2021; Yager, 2014). Many of these cultivars are trademarked, meaning a fee is paid to the developer in order to grow them (Pollock and Crassweller, 2021). Red Delicious is an interesting case in that, despite the name, it's been bred so many times for its beautiful quintessential apple look that it has lost much of its original flavor (Yager, 2014). It is a hardy apple, however, and travels well, so it is sold in many retail food markets across the United States (Pollock and Crassweller, 2021; Yager, 2014).

Once produce is picked, much of it is immediately brought in from the field and washed in cold pressurized water, cooling it quickly and substantially, to halt the ripening process (Zhang, 2018). Produce is sorted and inspected visually before continuing along an automated line where it will be packaged for shipping

ROSEMARY E. TROUT AND JOSEPH J. TROUT

(Zhang, 2018; Sinha, 2012). Some vegetables such as small "baby" carrots are processed by extruding larger carrots through dies that are the desired size of a baby carrot and washed again as a quality measure to lower oxygen exposure and to preserve the lipid-soluble beta-carotene from becoming lighter in color and slightly mottled during transit.

Biodynamic farming methods include relying on the sun and seasons, planting various food crops interspersed with flowers and non-food crops, no fencing, and no synthetic fertilizers, pesticides, or herbicides. To the unknowing, these farms may seem disordered and messy, but they are intended to be naturally balanced (Biodynamic Association, 2021; Woese et al., 1997).

Physical and Chemical Differences between Artisanal and Mass-Produced Fruits and Vegetables

Often, conventional fruits and vegetables are picked when immature and shipped to various states. Locally grown produce is more likely to be harvested when ripe and sold locally. *Local* is another term that is not well defined. Some artisan farmers use the range of within 500 miles, some less so (Cirne et al., 2019). They are also more likely to be organic or grown with organic farming methods but without the formal certification required to use the organic seal on or in reference to the produce. There is much variation between physical and chemical differences between conventional and artisanally produced fruits and vegetables. Post-harvest handlers think in terms of describing the physical and chemical differences in fruits and vegetables according to weight, sugar content, color, texture, and firmness, whereas consumers and food marketers will describe them in terms of consumer acceptance and use terms such as *good, superior,* and *fair.* Physical and chemical attributes change as the produce moves through the supply chain from grower to handler to marketer to consumer (Sinha, 2012).

Harvest and Post-production

While life on earth has been around for an estimated 4 billion years, flowering plants that yield fruits and other vegetative varieties that yield vegetables are 200 million years old. During this time, they have developed protective chemicals for survival. Some of these compounds are spicy, like capsaicinoids in chili peppers and thiocyanates in wasabi, horseradish, and mustard. Others are bitter, like theobromine, and still others astringent like tannins. These chemicals are expressed in produce from conventionally grown and small-scale farms. Advancements in post-harvest treatments have extended the shelf-life of produce in non-traditional ways and may alter the composition of produce (McGee, 2004).

Conventionally grown produce is harvested manually or mechanically with picking devices such as mechanical harvesters; an example is tree shaking so that immature fruit falls to the ground (Zhang, 2018). Some produce undergoes post-harvest heat treatments to lower microbial load or reduce insect infestations. This is done by hot water treatments, steam, or hot air circulation. Conventional produce may also be rinsed with a chlorine solution. Post-harvest chlorine usage conjures environmental concerns.

The shelf life of conventionally grown fruits and vegetables may be extended by the use of edible films and coatings (Zhang, 2018; Ciolacu et al., 2014; Díaz-Montes and Castro-Muñoz, 2021). Edible coatings are continuous biopolymeric matrices formed as films and directly applied on the exterior surface of fresh fruits. Edible coatings are solutions and emulsions made from proteins, lipids, and polysaccharides and are applied on produce surfaces by different mechanical procedures, such as dipping, spraying, brushing, or electrostatic deposition. Edible films are thin films that act as a barrier material to control moisture and exchange of gases such as oxygen, carbon dioxide, and ethylene. They also protect flavor compounds and minimize aroma exchange between proximate foods in storage. Edible films can also be used as wraps or pouches for fruits. Fruits and vegetables continue to respire after harvest, which means that there are changes to carbohydrates, organic acids, water content, texture, and color that continue until the fruit or vegetable is deemed rotten (Ciolacu et al., 2014; Díaz-Montes and Castro-Muñoz, 2021). Extending the time before these changes affect these chemical and physical attributes is beneficial to the growers and continues to improve the crop yield over time. Examples of conventionally grown fruits and vegetables are lemons, avocados, cucumbers, rutabagas, apples, mangoes, and papayas (Ciolacu et al., 2014; Díaz-Montes and Castro-Muñoz, 2021).

Artisanal or heirloom fruits and vegetables generally do not have edible films or coatings. Prior to the late 19th century, fruits and vegetables were preserved by canning, dehydration, salting, or the making of jams and jellies. So while not in their original form, fruits and vegetables harvested in one season could be kept and consumed in another season (Zhang, 2018).

Conventional produce is graded according to size, color, and weight. This is done by machine (colorimeters, light reflection techniques). Grading is done internally by the company producing the fruits and vegetables (Khoje and Bodhe, 2015). Artisanal produce may not be graded.

Storage

Produce from conventional farming systems typically requires transportation from the farm to the point of sale. Advancements in post-harvest technology have made produce more stable during shipping. This requires storage immediately after harvest as well as during shipping and prior to sale. Storage underscores the importance of preserving the sensory attributes, nutritional density, and shelf life of conventionally grown produce (Sinha, 2012). In many countries, produce is managed in open storage units with proper ventilation. Examples of different storage methods include refrigeration, controlled-atmosphere storage, modified atmosphere, hypobaric, storage at low temperatures with intermittent warming, and modified atmospheric packaging. There is a very wide variety of ranges for storage temperatures, relative humidity, and storage methods for produce.

Artisanal, small-scale farming uses minimal storage and relies on local purchasers on the farm, shipping directly to local retail stores, or ventilated storage of produce in boxes for direct shipping to members of community-supported agriculture (CSA). Conventionally grown produce may travel hundreds or thousands of miles before retail sales. Local, artisanal produce has shorter supply chains (Macieira et al., 2021).

Chemical Composition

There is scientific evidence that sustainably artisanal organically grown crops contain higher total phenolic content. In the study by Worthington (1998), artisanal Chinese cabbage stalks contained higher total phenolic content compared to their conventionally grown counterparts, and artisanal organic Chinese cabbage leaves contained higher protein percentages than those conventionally grown. Asami et al. (2003) studied total phenolics of conventionally, organically, and sustainably grown artisan marionberries, strawberries, and corn that went through freezing, freeze-drying, and air-drying processes. They found that for each processing method all three products—marionberries, strawberries, and corn—had higher total phenolic content, as well as higher ascorbic acid content, when compared to conventional and organic artisan farming systems. In the study by Renaud et al. (2014), various broccoli cultivars were analyzed according to whether they were grown conventionally versus organically, with genotype, region, and growth season. In this study organic versus conventional management systems contributed the smallest source of nutritional variation of glucosinolates, tocopherols, and carotenoids compared to genotype, region, and season. Neoglucobrassicin, α-tocopherol, zeaxanthin, and β-carotene were found in higher concentrations across cultivars, considering genotypes and locations. In the study by Meyer and Adam (2007), organic artisanal broccoli and red cabbage both contained significantly higher amounts of glucobrassicin than the same cultivars grown conventionally. The study by Hallmann and Rembiałkowska (2012), conducted in Poland, showed that organic bell peppers contained statistically significantly higher concentrations of ascorbic acid, total carotenoids, total phenolics, and flavonoids as compared to conventionally grown bell peppers. In the study by Aldrich et al. (2010), organically grown small production tomatoes of various cultivars shower higher overall ascorbic acid compared to the same cultivars grown with conventional methods. Amodio et al. (2007) showed that the ascorbic acid and total phenolic contents were higher in organic kiwifruits, which also presented a higher antioxidant activity, with higher percentages of total Ca and Mg.

Importance and Desirability of Heirloom Varieties

Heirloom and *heritage* are other terms, similar to *artisan*, that are not defined according to the US Food and Drug Administration or trade associations. In common parlance, *heirloom* more often refers to varieties of food crops that existed prior to commercialization, and *heritage* is applied to farm animals raised as food. Crop heritage is a growing global phenomenon whereby people conceive of change to agriculture in terms of potential information losses and promote means to safeguard what remains for future generations (Wincott, 2018).

Some heritage varieties are known for their rich and appealing flavor, as opposed to hardiness and yield. There are estimated to be over 1,000 seed banks in the world that store food seed for posterity and in the event of international turmoil of various sorts. The Svalbard International Seed Vault, also known as the Doomsday Vault, opened for storage in February 2008. It is located in the side of a mountain in Longyearbyen, Norway. It is in this location with the hope that it can withstand natural disasters as well as potential war, or human-made disasters. Other protected seed banks include the following:

- Berry Botanic Garden (Portland, Oregon, USA): Seeds from endangered plants of the Pacific Northwest
- International Center for Tropical Agriculture (Coli, Colombia): Cassava, forages, beans
- International Potato Center (Lima, Peru): Potatoes
- International Institute for Tropical Agriculture (Ibadan, Nigeria): Groundnut, cowpea, soybean, yam
- International Rice Research Institute (Los Banos, Philippines): Rice

Future Prospects

In addition to a definition for the term *artisanal* with regard to fruits and vegetables, the terms *local* and *organic* should be defined. Also indicated in the literature surveyed by these authors, cultivars, location, soil conditions, and seasonality should be included in any definition of the term *artisanal* (Aldrich et al., 2010). Exploration of hydroponics, urban gardening, aquaponics, and other farming methods as well as produce delivery systems beyond CSAs, retail food outlets, and farmers markets are also being considered.

References

Aldrich, H. T., K. Salandanan, P. Kendall, M. Bunning, F. Stonaker, P. Külen, and C. Stushnff. "Cultivar Choice Provides Options for Local Production of Organic and Conventionally Produced Tomatoes with Higher Quality and Antioxidant Content." *Journal of the Science of Food and Agriculture* 90 (2010): 2548–2555. https://doi.org/10.1002/jsfa.4116.

Amodio, M. L., G. Colelli, J. K. Hasey, and A. A. Kader. "A Comparative Study of Composition and Postharvest Performance of Organically and Conventionally Grown Kiwifruits." *Journal of the Science of Food and Agriculture* 87 (2007): 1228–1236. https://doi.org/10.1002/jsfa.2820.

Asami, D. K., Y. J. Hong, D. M. Barrett, and A. E. Mitchell. "Comparison of the Total Phenolic and Ascorbic Acid Content of Freeze-Dried and Air-Dried Marionberry, Strawberry, and Corn Grown Using Conventional, Organic, and Sustainable Agricultural Practices." *Journal of Agricultural and Food Chemistry* 51 (2003): 1237–1241. https://doi.org/10.1021/jf020635c.

Biodynamic Association. "Biodynamic Principles and Practices." December 8, 2021. https://www.biodynamics.com/biodynamic-princip les-and-practices.

Bourn, D., and J. Prescott. "A Comparison of the Nutritional Value, Sensory Qualities, and Food Safety of Organically and Conventionally Produced Foods." *Critical Reviews in Food Science and Nutrition* 42 (2002): 1–34. https:// doi.org/10.1080/10408690290825439.

Ciolacu, L., A. Nicolau, and J. Hoorfar. "Edible Coatings for Fresh and Minimally Processed Fruits and Vegetables." In *Global Safety of Fresh Produce*, edited by J. Hoorfar, 233–244. Amsterdam: Elsevier, 2014. https:// doi.org/10.1533/9781782420279.3.233.

Cirne, C. T., M. H. Tunick, and R. E. Trout. "The Chemical and Attitudinal Differences Between Commercial and Artisanal Products." *NPJ Science of Food* 3 (2019): 19. https://doi.org/10.1038/s41538-019-0053-9.

Cope, S. *Small Batch: Pickles, Cheese, Chocolate, Spirits, and the Return of Artisanal Foods*. London: Rowman & Littlefield, 2014.

Díaz-Montes, E., and R. Castro-Muñoz. "Edible Films and Coatings as Food-Quality Preservers: An Overview." *Foods* 10 (2021): 249. https://doi. org/10.3390/foods10020249.

Hallmann, E., and E. Rembiałkowska. "Characterisation of Antioxidant Compounds in Sweet Bell Pepper (*Capsicum annuum* L.) under Organic and Conventional Growing Systems." *Journal of the Science of Food and Agriculture* 92 (2012): 2409–2415. https://doi.org/10.1002/jsfa.5624.

Hartman, L. R. "Artisanal Foods Increase in Popularity." Food Processing, June 29, 2015. https://www.foodprocessing.com/articles/2015/artisanal-foods-increase-in-popularity/.

Heimler, D., P. Vignolini, P. Arfaioli, L. Isolani, and A. Romani. "Conventional, Organic and Biodynamic Farming: Differences in Polyphenol Content and Antioxidant Activity of Batavia Lettuce." *Journal of the Science of Food and Agriculture* 92 (2012): 551–556. https://doi.org/10.1002/jsfa.4605.

Khoje, S. A., and S. K. Bodhe. "A Comprehensive Survey of Fruit Grading Systems for Tropical Fruits of Maharashtra." *Critical Reviews in Food Science and Nutrition* 55 (2015): 1658–1671. https://doi.org/10.1080/10408 398.2012.698662.

Macieira, A., J. Barbosa, and P. Teixeira. "Food Safety in Local Farming of Fruits and Vegetables." *International Journal of Environmental Research and Public Health* 18 (2021): 9733. https://doi.org/10.3390/ijerph18189733.

Mapes, G. "Marketing Elite Authenticity: Tradition and Terroir in Artisanal Food Discourse." *Discourse Context Media* 34 (2020): 100328. https://doi. org/10.1016/j.dcm.2019.100328.

McGee, H. *On Food and Cooking: The Science and Lore of the Kitchen.* New York: Scribner, 2004.

Meyer, M., and S. T. Adam. "Comparison of Glucosinolate Levels in Commercial Broccoli and Red Cabbage from Conventional and Ecological Farming." *European Food Research and Technology* 226 (2007): 1429–1437. https://doi.org/10.1007/s00217-007-0674-0.

Pollock, R., and R. Crassweller. "Why All the New Apple Varieties?" Penn State Extension, February 4, 2021. https://extension.psu.edu/why-all-the-new-apple-varieties.

Renaud, E. N., E. T. L. van Bueren, J. R. Myers, M. J. Paulo, F. A. van Eeuwijk, N. Zhu, and J. A. Juvik. "Variation in Broccoli Cultivar Phytochemical Content under Organic and Conventional Management Systems: Implications in Breeding for Nutrition." *PLOS One* 9 (2014): e95683. https://doi.org/10.1371/journal.pone.0095683.z.

Robinson, J. C., and V. G. Saúco. *Bananas and Plantains.* 2nd ed. Wallingford, UK: CABI, 2011.

Sánchez, S. A., A. D. Morales, J. C. Castillas, C. A. Martínez, and A. Z. Meza. "Proposal for an Automated Greenhouse to Optimize the Growth of Hydroponic Vegetables with High Nutritional Content in the Context of Smart Cities." *IOP Conference Series: Materials Science and Engineering* 1154 (2021): 12012. https://doi.org/10.1088/1757-899X/1154/1/012012

Sinha, N. K. *Handbook of Fruits and Fruit Processing.* 2nd ed. New York: Wiley, 2012.

US Department of Agriculture. "USDA Coexistence Fact Sheets: Conventional Farming." February 2015. https://www.usda.gov/sites/default/files/documents/coexistence-conventional-farming-factsheet.pdf.

Wincott, A. "Treasure in the Vault: The Guardianship of "Heritage" Seeds, Fruit and Vegetables." *International Journal of Cultural Studies* 21 (2018): 627–642. https://doi.org/10.1177/1367877917733541.

Woese, K., D. Lange, C. Boess, and K. Bögl. "A Comparison of Organically and Conventionally Grown Foods—Results of a Review of the Relevant Literature." *Journal of the Science of Food and Agriculture* 74 (1997): 281–293. https://doi.org/10.1002/(SICI)1097-0010(199707)74:3<281::AID-JSFA 794>3.0.CO;2-Z.

Worthington, V. "Effect of Agricultural Methods on Nutritional Quality: A Comparison of Organic with Conventional Crops." *Alternative Therapies in Health and Medicine* 4 (1998): 58–69.

Yager, S. "The Awful Reign of the Red Delicious." *The Atlantic*, September 10, 2014. https://www.theatlantic.com/health/archive/2014/09/the-evil-reign-of-the-red-delicious/379892/.

Zhang, Q. *Automation in Tree Fruit Production: Principles and Practice.* Wallingford, UK: CABI, 2018. https://doi.org/10.1079/9781780648 507.0000.

.